FROM FREQUENCY TO
TIME-AVERAGE-FREQUENCY

FROM FREQUENCY TO TIME-AVERAGE-FREQUENCY

A Paradigm Shift in the Design of Electronic Systems

LIMING XIU

IEEE Press Series on Microelectronic Systems

WILEY

For general information on our other products and services or for technical support, please contact our
Customer Care Department within the United States at (800) 762-2974, outside the United States at
(317) 572-3993 or fax (317) 572-4002.

Wiley also publishes its books in a variety of electronic formats. Some content that appears in print may
not be available in electronic formats. For more information about Wiley products, visit our web site at
www.wiley.com.

Library of Congress Cataloging-in-Publication Data is available.

ISBN: 978-1-110-02732-4

Printed in the United States of America

CONTENTS

FOREWORD

DR. AHMAD BAHAI
Chief Technology Officer
Texas Instruments Inc.
Stanford University

Unprecedented innovations in semiconductor technology and manufacturing over last several decades have prompted an impressive growth in high-tech industry, which has immensely influenced our world. Today, electronics has permeated many aspects of our daily lives such as communication, transportation, health, and even our social interactions. A unique combination of efficiency, intelligence, form factor, and affordability of integrated circuits promises to address some of most daunting social and environmental challenges of modern world such as energy, health, and pollution. As more people and things are increasingly connected to each other and to the cloud of massive data and information, we will undoubtedly witness an ever-increasing rate of innovative ideas and applications enabled by semiconductor technology in years to come.

Almost every modern electronic circuit and system requires a timing reference. Reliable generation, transmission, reception, and conditioning and processing of data need timing references as well as some level of synchronization even in asynchronous systems! As the data rate is exponentially growing, the need for accurate timing reference exacerbates. Many advance high-speed data acquisition, interface, and processor-integrated circuits require a clock accuracy of a few femtoseconds (10^{-15})! In addition to a high-accuracy clock reference, generation of different frequencies and distribution of high-speed timing signal in a compact noisy path across the chip or the board are critical and challenging in high-speed analog and digital signal processing. Since the RMS jitter is inversely proportional to Quality Factor of the circuit, a combination of high-performance clock reference, low-noise circuit architecture, high-performance device, and careful layout is critical for a timing solution. The complexity and cost of clock generation and distribution amount to almost one-third of total board cost in many high-performance electronic boards and modules.

The infrastructure of modern world depends on high-performance clock and timing. High-speed wireless and wired communication networks which are the backbone of Internet and connectivity, power distribution systems, security systems, industrial metrology, scientific research systems, and many other applications increasingly rely on highly accurate and stable timing references.

We have come a long way from using celestial bodies' rotation and revolution as timing references. Modern timing references can provide accuracy, precision, and stability needed for high-performance circuits. Crystal oscillators and its high-stability variations such as VCXO, TCXOs and OCXO, MEMS-based clock references, GPS timing reference, and recently developed chip-scale atomic clocks (CSAC) can provide low jitter and phase noise integrated timing references. Also, advances in phase lock loop, injection-locked oscillators, and frequency synthesizer's theory and techniques such as hybrid and cascade architectures facilitate generation of accurate and stable timing reference at various frequencies as well as recovery and jitter cleaning of timing signal.

As the complexity of electronic systems grows the need for higher performance, robust timing references are becoming more pivotal and challenging. Liming Xiu offers a timely and comprehensive analysis of fundamentals and new approaches in timing and frequency synchronization techniques. The author's theoretical and practical experience through many years of advance R&D in Texas Instruments have been instrumental to providing a useful reference for one of the most challenging topics of today's technology.

I commend Liming for his insightful and articulate text which goes beyond conventional approaches by exposing the readers to many novel ideas and inspiring them to think out of the box.

It's about time.

PREFACE

A PARADIGM SHIFT

The word *paradigm* is defined in the dictionary as "a framework containing the basic assumptions, ways of thinking, and methodology that are commonly accepted by members of a scientific community." In his influential book *The Structure of Scientific Revolutions*, published in 1962, Thomas Kuhn used the term *paradigm shift* to indicate a change in the basic assumptions (the *paradigms*) within the ruling *theory* of *science*. Today, the term paradigm shift is used widely, both in scientific and nonscientific communities, to describe a profound change in a fundamental model or perception of events.

Ever since the clock concept was introduced into microelectronic system design many decades ago, it was assumed that all the cycles in a clock pulse train have to be equal in their lengths (a rigorous clock signal). One reason that this form of clock signal has dominated microelectronic system design for a long time is that, in the past, the requirement for IC clocking was mostly straightforward. A clock signal with a fixed rate was sufficient for most systems. However, the complexity of future systems changes the game. Low-power operation, low electromagnetic radiation, synchronization among networked devices (e.g., Internet of Things), complex data communication schemes, etc., all require a clock signal that is flexible.

Another reason behind the dominance of this style of rigorous clock is that time, which shows its existence and its flow indirectly through the use of a clock pulse train, is not a physical entity that can be controlled and observed directly. Thus, creating a flexible clock is an inherently difficult task. It demands effort beyond simply playing with various techniques at the circuit level. Philosophically it requires an adjustment, at a fundamental level, in our thinking about the way of clocking microelectronic systems. The "anomaly" in this case is a new perspective on the concept of clock frequency. In this line of argument, the materials presented in this book induce a paradigm shift in the field of microelectronic system design.

IN ELECTRICAL WORLD WE ONLY DEAL WITH TWO THINGS: LEVEL AND TIME

Although there are numerous different types of microelectronic devices and systems supporting the daily operation of our society, we only deal with two things when designing such devices and systems: level and time. Microelectronic devices and systems perform their magic by creating a variety of events that occur inside the silicon chip in a predetermined order. The purpose of such events is to essentially specify "what happens at when." In the process of creating those events, we need "level" to represent "what" and "time" to describe "when."

In describing "what," there are two approaches to implementation: (1) the analog way and (2) the digital way. The analog method uses proportional relationships to describe the physical world. (*Physical world*: It is the sum of all the stuff around us; you can see it, touch it, taste it, hear it, or smell it. And these five senses are based on the *proportional* relationship.) By contrast, the digital approach employs a binary system (i.e., on/off) to represent information. It is the natural language for performing computation using microelectronic devices. In the past several decades of silicon chip design, the task of describing "what" has been studied in great depth. Perhaps, it is fair to say that it is a mature art now.

However, we have not been as creative in dealing with "when." Historically, we were fixed in the belief that any clock cycle has to be exactly the same as any other cycle. Hence, we restrained our hand at making the clockwork for the electrical world. Since "time" is half of the story in "what happens at when," it can impact the microelectronic system's overall information processing efficiency in great deal. A small step change in the fundamental level (the anomaly) can produce a profound influence on upper level structures. This flow of thought is reflected in the development from the ideas of Chapters 1–4 to their applications in Chapters 5–7.

INTERNET OF THINGS AND THE CLOCK

The Internet of Things (IoT) is a growing network of everyday objects, from industrial machines to consumer goods, which can share information and complete tasks without human interference. It comprises three key components: (1) the things themselves, (2) the communication networks connecting them, and (3) the computing systems that make use of the data flowing among the things. IoT is the catalyst for new business growths across multiple industries, including industrial, medical, consumer, and automotive. The semiconductor industry, which provides chips designed for various IoT applications, is the enabler of this IoT trend. Designers of microelectronic products for IoT applications, however, face several unique challenges. The three most noticeable ones are the ultralow-power challenge, the ultralow-cost challenge, and the miniaturization challenge.

As said, IoT is a network of many things. It implies that the key in IoT is the "connection." For things in IoT to connect, it requires the establishment of a "common view of time" among the things. In other words, time synchronization of the network plays a major role in IoT. There are two essential pieces for establishing this synchronization: Each thing must have its own time (frequency) source to control its internal operation and there must exist a communication protocol agreed by all the things to establish and maintain the "common view of time." The design of this communication protocol depends heavily on the quality of the time (frequency) source. In IoT's harsh design environment of ultralow power, extreme small size, and ultralow cost, building a good time (frequency) source for each thing is an extremely challenging task. It requires innovations on clocking. Chapter 5 of this book provides some innovative options to meet this challenge.

CLOCK IS ENABLER FOR SYSTEM-LEVEL INNOVATION

Viewing from a high level, there are four fundamental technologies supporting the entire IC design business: processor technology, memory technology, analog/RF technology, and clock technology. In the past several decades, a tremendous amount of effort has been spent on the development of the first three technologies. Clock technology falls behind in this race. One of the key reasons for this is that, as mentioned before, clock technology deals with a special entity: time. It is neither directly observable nor directly controllable. The circuit designer can only play with it indirectly, through voltage and/or current. This lag, however, provides us an opportunity to make significant progress. It is a battleground for new ideas. It is a potential birthplace for great inventions. It is one of the enablers for system-level innovation. Chapters 5–7 are the first round of effort in this direction.

WHAT IS NEW ON CLOCK? FLEXIBILITY VERSUS SPECTRUM PURITY

When the term *flexible clock* is used, it refers to a clock signal whose frequency can be (1) arbitrarily set (within a small frequency granularity, similar to the way that voltage level can be arbitrarily reached within a quantization resolution) and (2) changed quickly (similar to the way that voltage level can make transition quickly). Preferably, these two features shall be achieved simultaneously and be available to the clock user at a reasonable cost.

A rigorous clock has the characteristic of high spectrum purity, which is beneficial to certain applications such as functioning as a carrier in wireless communication and as the driving clock for analog-to-digital converter. There are, however, many more applications wherein spectrum purity is not of high concern. Instead, a clock signal possessing the capability of small frequency granularity and fast frequency switching is more useful. Therefore, there is a crucial trade-off to be made when an IC design problem is investigated. In the past, a clock of high spectrum purity was the undeniable winner. However, for future microelectronic system design, this is not necessarily always the case. Chapters 5 and 6 of this book demonstrate that a flexible clock is more cost-effective in solving many emerging problems in modern applications.

CLOCK IS NOT PLL; IT IS MUCH BIGGER

Within the community of IC design professionals, a popular view is that IC clocking is just the PLL (phase-locked loop) design. This is far from the truth. PLL design is just one piece of a big puzzle. The PLL specializes in generating the clock pulse train. There are, however, many other aspects to the clock, including the task of delivering a clock signal, logically and physically, to all the areas that need it. Another important task is the correct use of the clock signal once it actually reaches the destinations (i.e., to drive the cells). This work is important because it can cause system failure if certain conditions are not satisfied (i.e., the setup and hold checks). Moreover, as a signal bearing highly concentrated energy at a particular frequency, the clock is a danger aggressor capable of doing serious damage to other signals around it. Thus, care must be given to avoid this from happening. Furthermore, the clock network consumes the largest percentage of overall chip power consumption. The reduction of power usage is heavily dependent on how the clock is used. Chapter 2 of this book helps the reader establish an appropriate appreciation of the clock: clock is not simply the PLL; it is a much bigger topic.

"JITTERY" CLOCK IS NOT NECESSARILY A BAD THING

Among clock circuit designers and clock signal users, jitter is always a bad thing. The essence of a clock pulse train is to create a series of "moments in the flow of time" by utilizing the mechanism of the "voltage-level-crossing-a-threshold." The resulting moments are used as the reference points for other events happening inside the microelectronic system. Therefore, the requirement on those moments is that their locations-in-time must be predictable and precise. Jitter is a parameter measuring this quality. Thus, a jittery clock is undesirable since it reduces the effectiveness of the clock in coordinating other events. However, jitter is not without any use. An obvious example of its applicability is that jitter in a clock signal can help reduce its electromagnetic radiation since it spreads the clock energy.

The not-so-obvious, and more valuable, use of a "jittery" clock is to trade the irregularity in the moment with the flexibility. As mentioned, the flexibility associated with a clock signal refers to its capability of fine frequency resolution and fast frequency switching. When used with care of this irregularity in the moment, a clock signal can be made flexible by intentionally introducing "controlled jitter" into it. This capability is important for certain applications, such as in adaptive clock generation and in clock data recovery. Indeed, it outweighs the requirement on the clock's spectrum purity in such applications. Hence, a jittery clock is not necessarily a bad thing. An example is provided in Chapter 5 to support this unconventional view. It also serves as an example of "thinking out of the box."

THE POWER OF IDEA

Many times in our history, the power of an idea has changed the landscape of our civilization. Such ideas include liberty, democracy, Romanticism, Confucianism, Marxism, Zionism, among others. Each of those great ideas led to a profound movement that changed the way we live. In science and technology, the latest example of such an idea would be Einstein's theory of relativity. It links the space and time together, resulting in a thing called space-time. This breakthrough idea, which was regarded as a ridiculous one by most people when it made its debut, is proven to be one of the greatest in human history. This idea is an "anomaly" that later leads to a great paradigm shift in science.

In my previous book *Nanometer Frequency Synthesis beyond Phase Locked Loop* (Wiley-IEEE Press, 2012, IEEE Press Series on Microelectronic Systems), a new perspective on clock frequency was introduced and its associated implementation technology was presented. This angle of using a clock pulse train is not aligned with the prevailing view. However, evidences show that it works. In many cases, it can do a better job with lower cost. After that book was published, I have often encountered the question of "what is it useful for?" from people of old fashion. While the material presented in that book focused on building the circuit at component level, this book will answer the question of how to use it in an upper level to create better systems. This book is the continuation in this route of new microelectronic system design methodology; it can be treated as Volume II of this series. Volume I (the 2012 book) teaches the technique of making a field programmable frequency generator (FPFG). With this FPFG available to chip designers, Volume II (this book) shows the ways of using it. The spirit of this series of books is innovation. The goal is to create cheaper, better, and faster products.

ACKNOWLEDGMENT

I received my bachelor's and first master's degrees from Tsinghua University (Beijing, China). Tsinghua's training prepared me to be a scientist with an innate curiosity on the structure of the world, with a strong intention to pursue the beauty of the universe. I earned my second master's degree from Texas A&M University (College Station, TX, USA). A&M's education turned me into an engineer with a desire to build and invent. I got my "PhD degree" from "University of TI" (Texas Instruments Inc., a.k.a. Training Institute). My "PhD adviser" is the collection of numerous TI engineers, including my wife (a long-time TIer).

The TI engineers fighting in the semiconductor front line taught me how to deal with real-world problems and, more importantly, equipped me with the capability to spot emerging problems. My "PhD dissertation" is the invention of Flying-Adder frequency synthesis architecture plus the book of *VLSI Circuit Design Methodology Demystified: A Conceptual Taxonomy* (quality controlled by the forewords written by TI CEO Rich Templeton and then TI CTO Hans Stork). I truly believe that my "PhD degree" acquired in this environment is competitive with the PhD degree issued from any top engineering school in the world. Novatek Microelectronics (Taiwan) is a place that provides me with room for doing some serious thinking, to sharpen my vision. The Novatek experience enabled me to compose my second book *Nanometer Frequency Synthesis Beyond the Phase-Locked Loop*. The warmth of Novatek people will be remembered for life. My 2-year tenure as vice president of IEEE Circuit and System Society is unique. It is beneficial to me in many ways. It has broadened my view from another direction.

This book would be impossible if any part of the aforementioned experience was missed. It is the result of knowledge-and-experience accumulation over three decades. This long process is enjoyable most of the time, but painful occasionally.

Thanks to all those who helped me in this journey. I also want to thank my editors in Wiley & IEEE, Mary Hatcher and Brady Chin, for their great help. Their efficiency is greatly admired by me. Their patience on me is deeply appreciated. Finally, I want to give special thanks to my dear wife Zhihong You (my Tsinghua, Texas A&M, and TI alumnus). She allowed me to spend time on this book in countless evenings and weekends while she virtually acted alone on raising our two daughters, and successfully sent the elder one to Yale University in 2014.

LIMING XIU
DALLAS, TX
2015

1

IMPORTANCE OF CLOCK SIGNAL IN MODERN MICROELECTRONIC SYSTEMS

1.1 CLOCK TECHNOLOGY: ONE OF THE FOUR FUNDAMENTAL TECHNOLOGIES IN IC DESIGN

Today's typical electronic systems contain millions of electrical signals. They make the system perform what it is designed to do. Among these, the most important one is the clock signal. From an operational perspective, the clock is the timekeeper of the electrical world in a chip/system. From a structural perspective, the clock generator is the heart of a chip; the clock pulse is the heart beat; the clock signal is the blood; and the clock distribution network is the vessel.

The timekeeper has played and is playing a critical role in human life. History shows that the progressive advancement of our civilization is only made possible by the steady refinement of the timekeeper: the clock [Fra11]. The same is true for electronic systems. The purpose of electronic systems is for processing information. The efficiency of performing this task is highly dependent on the time scale used. This time scale is controlled by the clock signal. It has two key aspects: its size (the absolute clock frequency) and its resolution (the capability of differentiating nearby frequencies; resolution can also be viewed as frequency granularity and/or time granularity). In addition, another characteristic is important in the electronic system: the speed at which the time scale can be switched from one to another (the clock frequency switching speed).

From the day of Robert Noyce [Noy61] and Jack Kirby's [Kil64] first integrated circuit in 1959 to today's systems of billions of transistors on a chip, the art of

From Frequency to Time-Average-Frequency: A Paradigm Shift in the Design of Electronic Systems, First Edition. Liming Xiu.
© 2015 The Institute of Electrical and Electronics Engineers, Inc. Published 2015 by John Wiley & Sons, Inc.

integrated circuit (IC) design can be roughly individualized into three key areas: *processor* technology, *memory* technology, and *analog* technology. Processor technology focuses its attention on how to build efficient circuits to process information. Using transistors to do logic and arithmetic operations with high efficiency is its highest priority. Memory technology is the study of storing information in a circuit. Its aim is to store and retrieve information in large amounts and at high speed. Analog technology squares its effort at circuits that interface electrical systems with humans (or the world of physical phenomena). Inside electronic systems, information is processed in binary fashion. Once outside, information is used by us in proportional style since our five senses are built upon proportional relationships. The analog circuit is the bridge in between. During the past several decades, advancements in these three circuit technologies have made today's electronic systems very powerful. However, the driver of these three technologies, the clock, has not seen fundamental breakthroughs. The time scale is not flexible: The available clock frequencies are limited and the switching between frequencies is slow. To improve the electronic system's information processing efficiency further, the next opportunity is with the method of clocking: (1) We need a flexible on-chip clock source and (2) and it needs to be available to chip designers at a reasonable cost. Now is the time for clock to be recognized as a technology, as illustrated in Figure 1.1.

There are four key challenges in the generation of a clock signal: high clock frequency, low noise, small frequency granularity (also loosely referred to as arbitrary frequency generation), and fast frequency switching (also loosely referred to as instantaneous frequency switching). The first two have been studied intensively by researchers. The last two have not drawn much attention. Another challenge lies in distributing the generated clock signal to all the places that require a clock. Clock distribution is a difficult problem both functionally and physically. From a functional perspective, a cell requiring a driving clock might need the clock signal to come from different sources in different operating modes. The logical path from a source to any destination (clock sink) is controlled by the selector, frequency divider, and gater, as illustrated on the left in Figure 1.2. These elements ensure that a clock sink sees the appropriate clock signal at the appropriate time. From a physical point of view,

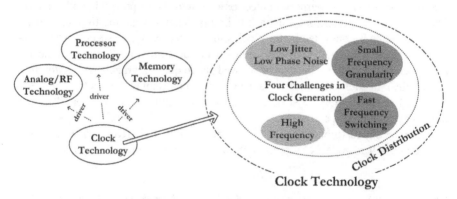

FIGURE 1.1   Clock as a technology.

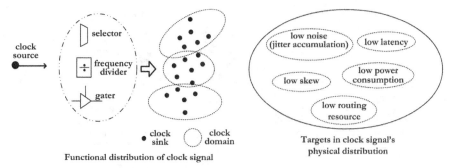

FIGURE 1.2 Clock distribution.

a clock signal from a source might need to be delivered to cells that are spread in a large area. The task of physical distribution has to be carried out with high fidelity (low noise, low skew), small delay (low latency), and low cost (in terms of routing resource and power consumption). These goals are presented on the right-hand side of Figure 1.2.

Clocking is an important and challenging topic in both academic research and engineering practice. IC clocking is closely tied to the two functions in modern chip design: communication and computation. IC clocking also plays an important role in determining the amount of energy consumed in performing these tasks. There are countless papers dedicated to its study in scientific journals and conference proceedings. It is perhaps the most studied topic in electrical engineering. There are also many books devoted to this area of study; most of them focus on the phase-locked loop (PLL) [Gar05, Bes07, Ega07, Raz03, Fri95]. For those reasons, it is fair to recognize IC clocking as one of the four fundamental circuit design technologies. This book is not focused purely on the PLL, which is only part of the clock story (that of clock signal generation). It addresses the IC clocking issue in a much larger scope: clock frequency, clock generation, clock distribution, and clock application. The essence of this book is to influence the landscape of IC design from the clocking side, starting from the fundamental concept of clock frequency.

1.2 CLOCK SIGNAL GENERATOR: THE KNOWLEDGE-AND-SKILL GAP BETWEEN ITS CREATOR AND ITS USER

When a clock signal is used in an electronic system, it involves two groups of engineers: designers of the clock generator and users of the clock signal. This scenario is illustrated in Figure 1.3. These two groups possess different knowledge-and-skill sets. The clock generator designer (usually a PLL designer) focuses his or her attention on creating a circuit that produces an electrical pulse train. The main interest in this task is the quality of the pulse train: the available frequency range, the granularity of its frequency, the frequency switching speed, and the amount of noise embedded in the pulses. The key skill required lies in the area of analog circuit design. Understanding the noise generation mechanism in semiconductor devices is also important.

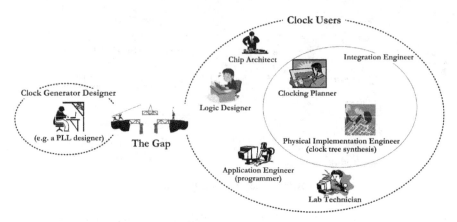

FIGURE 1.3   Clock signal generator: the gap between its creator and its users.

Clock users can be divided into several subgroups. The chip architect is responsible for designing the chip system. Communication between various parts of the chip, as well as the task of computation, is controlled by the clock signals. Thus, the chip architect must have a solid understanding of how the various clock signals are used to perform the computation and communication tasks in the chip. In some design cases, it is possible to have over 100 clock domains in a large system. All have to be carefully designed by the chip architect. A logic designer helps the chip architect realize the chip functional specifications using hardware description language [such as very high speed integrated circuit (VHSIC) Hardware Description Language (VHDL) and verilog] or higher level system languages. A large amount of simulation must be performed by the logic designer to ensure the correctness of the chip functionality. In such simulations, among perhaps millions of on-chip signals, the clock signal is the most studied one. When something unexpected happens in a simulation, the first signal that the designer turns his or her attention to is usually the clock signal.

Integration engineers receive the chip design in a words-and-diagrams description and turn it into a functioning system represented by metals and semiconductor devices. In this process, clock implementation is a crucial part. Where chip clocking is concerned, the clocking planners take the instructions of the chip architect and turn them into an implementable plan. In this task, the clocking planner needs not only to fulfill functional requirements but also to pay attention to a variety of chip testing concerns. The physical implementation engineer (also called a place-and-route engineer) takes the logical plan from the clocking planner and realizes it using metals and standard cells. This work is commonly termed clock tree synthesis. Finally, the application engineer needs to have a firm grip of the chip's clock structure so that the bare chip can be programmed to do its task.

The subjects dealt with by these two groups of engineers are significantly different. The clock circuit designer deals with transistors. Clock users deal with various entities such as functional blocks [e.g., Digital Signal Processor (DSP), Universal Serial Bus (USB), double data rate synchronous dynamic random-access memory (DDR)], standard cells, Hardware Description Language (HDL) coding and simulation, Static

Timing Analysis (STA) timing analysis, place and route, etc. They speak different languages and use different tools. Although the ultimate goal is to make the chip perform as designed, the immediate aims of each type of engineer are very different. The knowledge-and-skill sets required by those engineers are sophisticated enough that a large gap exists between the two groups. For example, a PLL designer oftentimes does not know (or does not have to care) how the chip architect or clocking planner would use his or her PLL. Sometimes, this is caused by a lack of knowledge of other fields. Other times, this could be due to the organizational boundaries of different groups (or different companies).

Within the world of clock circuit designers, the PLL has traditionally been the architecture of choice for the on-chip clock generator. The PLL is a beautiful blend of digital and analog circuits in one piece of hardware. From a given reference time scale, it can generate other time scales. However, due to its use of a *compare-then-correct* feedback mechanism, the choice of time scales that can be produced is limited (it is difficult to make frequency granularity small). Equally harsh is the problem that the change of time scale (frequency switching in PLLs) takes a very long time. Although the PLL has played a key role in making today's electronic systems magnificent, these two problems are limiting the chip architect's capability for creating further innovation at the chip level.

The root of these two problems is partially due to this gap. As a clock circuit designer, the goals of arbitrary frequency generation and fast frequency switching are difficult to achieve, especially simultaneously (in contrast, arbitrary voltage generation and fast voltage switching are easy to do). On the other hand, chip/system architects, from the day the clock signal is introduced into the field of chip design, have not asked the clock circuit designer about these two features since they know that it is difficult. As a result, the clock circuit designer does not have the motivation. The problems associated with these two point-of-views are cause-and-effect of each other: The system architect does not know that it can be done; the circuit designer does not know that it is needed. The goal of this book is to break this lock, to provide a vision that it can be done and it is useful.

This gap, when properly addressed, could be the birthplace for important innovations.

1.3   HOW IS SENSE-OF-TIME CREATED IN ELECTRICAL WORLD?

Today's typical electronic systems contain millions of electrical signals. Signals are the medium for carrying information among electronic devices (transistor, diode, capacitor, resistor, inductor, etc.). Without clearly defined signals, an electronic system cannot perform any useful function except being a heat generator that converts electrical energy into thermal energy. Any electrical signal can be described by using two and only two physical properties: level and time. This is depicted on the left-hand side of Figure 1.4. These two properties correspond to the two fundamental phenomena of the universe: the mass of materials and the flow-of-time. The mass of a material represents the strength of a signal; it is the number of electrons currently being processed by an operating device. It shows its impact through the voltage (or

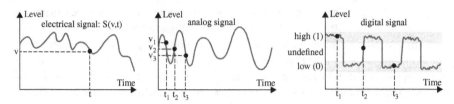

FIGURE 1.4 Electrical signal through level and time (left), analog signal (middle), and digital signal (right).

current) level. This level (the mass of the material) always varies with the flow-of-time. Therefore, level is only identifiable through a snapshot of the level taken at a particular moment in the flow-of-time. In the electrical world, similar to our social world, the flow-of-time must be quantified and each moment must be indexed through some mechanism. By the use of these two fundamental properties, an electrical signal can be expressed as $S(v, t)$, where v represents the level strength and t is the moment in the flow-of-time.

There are two ways to describe a signal level: analog and digital. In the analog approach, a proportional relationship is established between the intended information and the level. In other words, as shown in the middle drawing of Figure 1.4, every point on the Y-axis (level) is meaningful; the level is treated as a continuous variable. On the other hand, a binary relationship is the foundation of the digital methodology. As illustrated on the right-hand side of Figure 1.4, level is only distinguished by two regions: high and low. Other measurements of level have no meaning. Information is expressed only through these two values. For example, the snapshot of level at moment t_2 is invalid from the digital circuit designer's point of view.

The human body is naturally equipped with the capability of sensing the flow-of-time. To coordinate various events in our lives, however, the flow-of-time needs to be quantified and explicitly expressed in a certain way. Mechanical vibration serves this purpose as evidenced by the billions of watches and clocks used in today's society. Since level (the flow of electrons) is the only physical property that can be sensed by an electronic device, the flow-of-time in the electrical world must be quantified through this medium. In fact, in the electrical world, the flow-of-time shows its form by crossing a predefined threshold (as will be explained next). This is, in general, also a type of vibration—electrical oscillation.

In the electrical world, at the device level, the basic building elements are the transistor, diode, resistor, capacitor, inductor, etc. All these devices interact with each other through voltage and current (more precisely, through electrons). By manipulating the voltage/current magnitude, information is processed. Behind the scenes, the supporting mechanism is the fact that voltage/current magnitude is proportional to the number of electrons flowing through these devices. At the circuit level, information can be treated in two different ways: digital and analog. A digital circuit uses two states—on and off—represented by 0 and 1. In an analog circuit, information is established by a proportional relationship. At the functional level, information is collected by a sensing circuit, manipulated by a processing circuit (amplification/

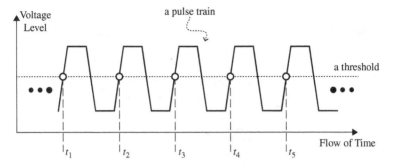

FIGURE 1.5   Using voltage-crossing-a-threshold and indexing to create the sense-of-time.

attenuation, screening/filtering, and logical/arithmetic operation), transformed between digital and analog formats by a converter circuit, and applied to an actuator circuit carrying out the action. At the architectural level, the information processing task can be classified as computation and communication.

From an operational perspective, the electronic system carries out its tasks through events. This is similar to our social world: Activities are composed of individual events, where tasks are accomplished through making events occur in an orderly fashion, that is, doing things in sequence. This requires a timekeeping mechanism— thus the invention of the clock (and later the watch). Similarly, for an electronic system to be useful, its events have to be organized by a timekeeper. By taking advantage of electrical devices' natural capability of differentiating voltage level, a marker system using voltage-crossing-a-threshold is created to indirectly mark the moments in time. The flow-of-time is created by indexing the markers. In this approach, a train of electrical pulses is established. During the process of the level oscillating between high and low states, a moment in the flow-of-time can be identified within each low-to-high (or high-to-low) transition. These moments are recorded and indexed by numbers, resulting in $t_1, t_2, t_3, \ldots$, as shown in Figure 1.5. These markers are the references for other functional events in the electronic system. This special train of pulses is termed a clock signal. Each individual pulse (a complete high-to-low or low-to-high cycle) is termed a clock cycle, which is identifiable by its index number. Mathematically, the most important requirement (actually the only requirement) imposed on this clock pulse train is that these moments of time $(t_1, t_2, t_3, \ldots)$ must be predictable. The accuracy in predicting them must be made as good as possible. The reason is that they are the markers used for coordinating other events. Any error in determining the locations of these moments can lead to a reduction in the effectiveness of organizing other activities. The larger the uncertainty of these locations (which is called clock noise or jitter), the less effective the pulse train is as a timing marker.

Clock frequency is defined as the number of pulses within a time window of one *second* (the second is defined from the atomic clock, based on the fundamental properties of nature [NIS09, Jon00]). To create a clock signal with a precise and stable frequency, the number of pulses has to be well controlled. This is however extremely difficult since electrical devices do not bear sense-of-time naturally. External help is

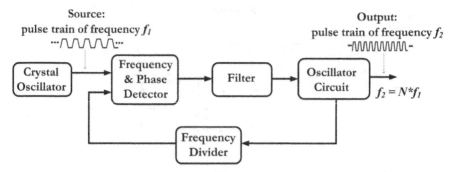

FIGURE 1.6   Using PLL to generate other frequencies from a reference frequency.

needed. Traditionally, a quartz crystal rock has been chosen to serve this purpose due to its high-precision mechanical vibration when a voltage is applied to it. An electrical supporting circuit is built around the crystal to convert the highly accurate mechanical vibrations to electrical pulses. The resulting circuit is called a crystal oscillator. For several decades, it has been the timing reference for almost every electronic system [Ger67, Vit88, Wal95].

Although highly accurate, the frequency of a crystal oscillator is fixed at a single value. As a result, the choice of timing marker is limited. This restrains the flexibility in the design of the electronic system. To cope with this difficulty, a special circuit called a phase-locked loop (PLL) has been created. Relying on negative feedback, the PLL utilizes the mechanism of *compare-then-correct* to create pulse trains of other frequencies using the fixed-frequency crystal oscillator as a reference. Its principal structure is depicted in Figure 1.6. The PLL is one of the foundational circuits in circuit design. It can be found in almost every modern chip. However, the solution is not perfect. Due to the feedback used, the PLL output has two problems: (1) The frequency cannot be arbitrarily generated (only some multiples of the reference frequency are available) and (2) switching from one frequency to another takes a long time. These two issues were not of high concern for system architects in the past when there were many other more immediate problems. However, to improve the electronic system's processing efficiency further, now is the time to reinvestigate them in depth.

From the clocking principle illustrated in Figure 1.5, it can be understood that the operable "time" inside an electronic system is realized by a frequency source (a pulse train at a certain frequency) followed by a counter. The counter, which is driven by this frequency source, records the index of each pulse as the "physical time" flows forward. Together, a clock source of a certain frequency and a counter make up a time scale. This mechanism is depicted on the left in Figure 1.7. As shown, the time inside the electronic system is represented by cycles that have elapsed since the start of counting. "Time (cycle)" refers to the integration of frequency over time. The value of the frequency can be derived by taking the first derivative of the time (cycle). Consequently, as shown on the right-hand side, noise on the frequency

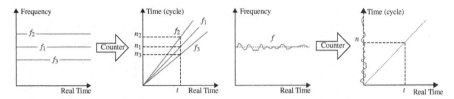

FIGURE 1.7 A time scale is a clock source of a fixed frequency plus a counter (left); noise in frequency source leads to error in timekeeping (right).

source (instability of its frequency) will be converted into error in time (cycle). Several important observations can be made regarding the relationship between time (cycle) and frequency:

1. **Time Granularity**   Time granularity is inversely proportional to the value of frequency. In other words, the higher the frequency, the finer the resolution in time will be.
2. **Frequency Granularity and Time Scale**   Having more frequencies available from a clock source leads to more choices of time scales. The number of choice for the time scale depends on the frequency granularity of the clock source.
3. **Change of Time Scale**   The switching speed from one time scale to another depends on how fast the frequency can be changed from one to another by the clock source.

The time scale is used to control the pace of the operation inside an electronic system. It indicates the effectiveness of a system to process information. On the other hand, the power consumed by an electronic system is directly related to the time scale chosen. For this reason, the time scale (and thus frequency) is one of the most important factors in electronic system design.

1.4   ALL MICROELECTRONIC SYSTEMS ARE FREQUENCY DRIVEN

An electronic system is used for processing information. Tasks included are collecting input information from an external environment, processing the collected information, and sending out the processed information to the external environment. These three functional blocks of input interface, digital processing, and output interface can be realized by individual chips or by a variety of functional blocks in a single SoC (System-on-a-Chip). Large or small, all electronic systems are driven by clocks of various frequencies. This fact is graphically depicted in Figure 1.8.

In the social world, information takes a continuous form. It exists in every moment in the flow-of-time. For information to be processed by an electronic system which does not itself have a sense-of-time, oftentimes information is sliced into pieces by a clocking mechanism (the action of sample and hold). Figure 1.9 shows a high-level

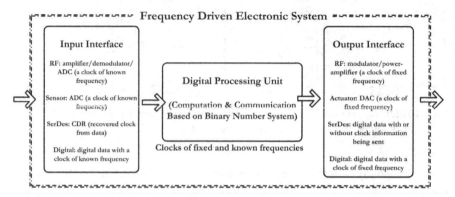

FIGURE 1.8   Electronic system is driven by clocks of various frequencies.

abstract view of how information is treated in an electronic system. After slicing, information is indexed and quantized through the use of numbers. In the input interface of Figure 1.8, for radio frequency (RF) and sensor applications, the slicing is done through an ADC (analog-to-digital converter). For SerDes and digital input cases, the input information is already in the sliced format (by previous processing units). On the output side, the processed information can be sent out in either sliced or continuous format. The processing unit between the input and output interfaces is used to manipulate the received information and consequently to create new information. This task can be classified into two subtasks of *computation* and *communication*. Computation includes logical and arithmetic operations. Communication is the action of moving data among interacting parties. The backbone of both computation and communication is the clock.

In modern systems, there can be hundreds of clock signals operating in a chip simultaneously to achieve sophisticated functions. Using the illustration in Figure 1.5, all clock signals are mathematically the same except one attribute for their identification: *the number of pulses existing in the time frame of one second*. This is termed the *clock frequency*. In the generic system of Figure 1.8, the clocks have unique frequencies that can differ from each other in large or small degrees. From this perspective, all electronic systems can be characterized as clock-driven systems. In a more abstract view, *an electronic system is a frequency-driven system.*

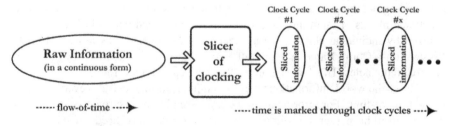

FIGURE 1.9   Information is sliced into pieces through clocking.

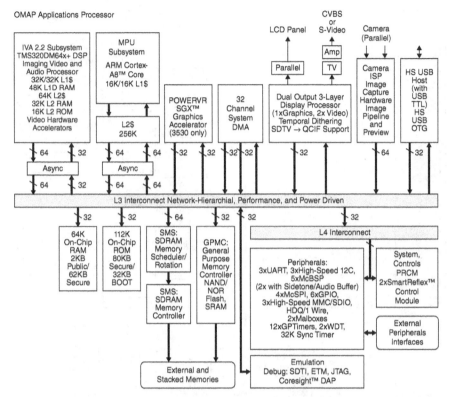

FIGURE 1.10 Video processor OMAP3530/25 system block diagram: a system of many functions integrated on-chip [Oma09]. Courtesy Texas Instruments Inc.

Figure 1.10 is an example of a very large scale integrated (VLSI) system. It is an Open Multimedia Application Platform (OMAP) processor which has several different types of processors on-chip [Oma09]. The core processor is ARM Cortex-A8; the DSP processor is TMS320DM64x+DSP. There is also a graphic accelerator POWER VR SGX. In addition to the processors, many other functions are integrated on-chip as shown, such as memory systems and various peripherals. All these functional blocks are running at their optimum frequencies, for example, AMR at 720 MHz, C64x DSP at 520 MHz, and USB at 48 MHz. The chip's frequency plan is very complex, as shown in Figure 1.11. A dedicated clock manager system is created to generate the various clock frequencies required. From this example, the message is very clear: An electronic system is frequency driven.

There are two trends in designing today and tomorrow's electronic systems: integrating more functional blocks into a system and increasing clock frequency (for faster information processing). These trends have an impact on almost all aspects of chip design. From the clocking point of view, they present two challenges. The first one is the efficiency in data communication between functional blocks that are often running at different frequencies: *the communication challenge in the environment of*

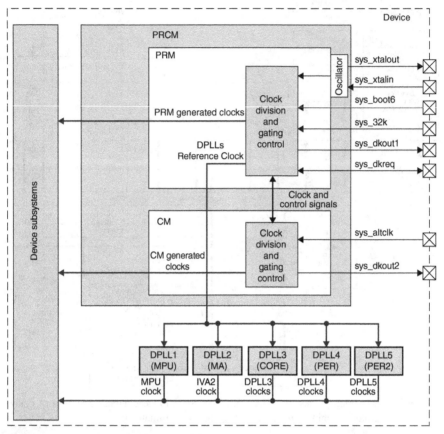

FIGURE 1.11   Video processor OMAP3530/25 clock manager for generating various clock signals of different frequencies [Oma09]. Courtesy Texas Instruments Inc.

heterogeneous clocking. As illustrated on the left in Figure 1.12, there could be many blocks in a large SoC system that perform different functions. Very likely, each will have its own optimum operating clock frequency (i.e., the best time scale suitable for the task it performs). When data exchange between blocks is required, the difference in their clock frequencies presents a major design challenge. This issue is typically handled by using a first-in-first-out (FIFO) memory in between for temporarily storing data. Depending on the magnitude of the frequency difference, the size (and thus the cost) of the memory could be very large. If an adaptive clock generator with the capability of arbitrary frequency generation and fast frequency switching is available, the severity of this problem can be mitigated. This issue will be discussed in more detail in later chapters.

The second problem is the *electromagnetic interference (EMI) associated with a high-frequency clock signal*. A high-frequency clock (fine time scale) enables fast information processing. It has, however, undesirable side effects on the electronic devices operating in its surrounding environment. EMI reduction requires a flexible

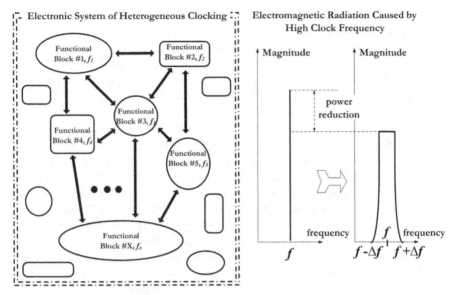

FIGURE 1.12   Multiple functional blocks of different operating frequencies leads to hetero-geneous clocking (left); strong EMI of high clock frequency calls for its reduction (right).

clock generator capable of generating many frequencies. As shown on the right-hand side of Figure 1.12, by spreading a single clock frequency to a group of frequencies, the power radiated by the clock generator can be reduced significantly. This issue will also be addressed in depth in a later chapter.

If we move from the chip level to a higher level, today's electronic system design presents an even more challenging problem in terms of timekeeping: *establishing a common view of time among networked systems*. As illustrated on the left in Figure 1.13, the majority of today's electronic devices are networked [e.g., the latest trend of Internet of Things (IoT)]. To create a temporal sequence in a networked environment so that events can occur in a sequential order, a common view of time needs to be established. The drawing on the right illustrates the method of generating *time* (time

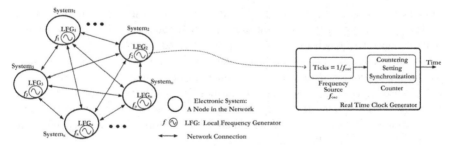

FIGURE 1.13   Future systems are networked: calling for a common view of time among all systems (left); method of generating *time* in each system (right).

here refers to time as used in our daily life: days, hours, minutes, seconds). A frequency source is used to generate ticks, which are then used to drive a counting mechanism. The counting mechanism is responsible for generating the real time locally in each system. It is also responsible for synchronizing its time with that of the rest of the systems in the network. Further, in some applications, it is synchronized with an external time reference, such as the UTC (Coordinated Universal Time) [All97].

In practice, each system in the network has its own clock generator [local frequency generator (LFG)] with a unique oscillation frequency. Usually, these frequencies are not matched perfectly among those devices in the networked system. As a consequence, the time flows among these systems will be out of synchronization (due to drift) even if they are initially set to the same time point. To coordinate time-sensitive events within the network, these timekeepers must be synchronized. The standard protocols NTP (Network Time Protocol) and PTP (Precision Time Protocol, IEEE 1588) are the two best-known examples for this purpose [Fer13]. NTP carries out its time synchronization task at the software level; the best result achieved is in the range of tens of milliseconds. PTP refines the synchronization mechanism and, consequently, the synchronization precision is improved to the microsecond range. Both NTP and PTP, however, address the time synchronization problem in algorithm and architecture level and do not consider hardware implementation in great detail. To improve the synchronization precision further (such as to the nanosecond range), an effective way is to refine the time and frequency granularity at the hardware level. When nanosecond network time synchronization is achieved, countless application possibilities will emerge that can significantly improve the network's information processing efficiency.

In summary, all electronic systems are frequency driven. Modern electronic system design calls for a new clock source that is flexible in its ability to generate frequency. This is a new requirement commanded by the ever-increasing complexity of modern systems. It is a new challenge. It is also a new opportunity.

1.5 A NEW KID IN TOWN: THE CLOCK ARCHITECT

The modern large SoC has passed the milestone of one billion transistors on a chip. Designing such a complex entity requires a group of highly skilled professionals. Among them, without question, the chip architect plays a leading role. His or her job description includes following the market trend, understanding customer requirements, being familiar with competitors' products, and designing a product that is appropriately aligned with his or her company's marketing strategy. His or her must-have knowledge and skill are computer architecture and communication protocol. He or she is also required to be familiar with various IPs (Intellectual Property) available on the market. His or her primary goal is to design a product that is computation and communication efficient (using minimum energy to perform a given task) and, more importantly, is right for the market.

When the target product is gradually moved from the design stage to the implementation stage, however, the knowledge and skill of a chip architect are probably no longer adequate. The knowledge and skill needed at this stage are related to electrical engineering (rather than computer engineering). The focus is shifted from "defining what we want in this chip" to "how to realize them by creating various events using on/off switching activities." Since the clock signal is the conductor of this symphony of events, this signal must be treated with extreme care. Further, since the creation of a clock signal is closely tied to the use of transistors and metals, the clock signal designer must have expertise in transistor level circuit design. This leads to a new type of IC design professional: the clock architect.

The clock architect is neither a chip architect nor a PLL designer. He or she must be able to use the clock signal to organize the large amount of events occurring in the chip. Given the tasks demanded from the chip architect, he or she must fulfill them using the least amount of resources in terms of power consumption and silicon area. By appropriately using the tools of clock frequency, clock frequency range, clock frequency granularity (resolution), and clock frequency switching speed, a good clock architect can make the difference between a successful product and a failed product.

The clock architect is not the integration engineer (e.g., the clock tree synthesizer) whose main focus is "how to implement the clock plan using silicon and metal". The clock architect stands at a higher level of planning the frequency game. In the past, when the clock generator could only produce a few frequencies and the switching between frequencies was slow, there was not much room for the clock architect to make a significant difference. However, with the small frequency granularity and fast frequency switching of today's clock generator, frequency planning becomes a major piece of design work that requires dedicated attention.

With the features of small frequency granularity and fast frequency switching becoming a reality, the clock architect is provided with the capability of programming the "clock frequency" of a clock generator just as the software programmer programs the instruction set of a processor. *In short, the clock architect deals with frequency.*

REFERENCES

[All97]   D. W. Allan, N. Ashby, and C. C. Hodge, "The science of timekeeping," Application Note 1289, Hewlett-Packard, 1997.

[Bes07]   R. Best, *Phase Locked Loops: Design, Simulation, and Applications*, 6th edn., McGraw-Hill Professional, July 2007.

[Ega07]   W. F. Egan, *Phase-Lock Basics*, 2nd edn., Wiley-IEEE Press, Hoboken, NJ, Nov. 2007.

[Fer13]   J. L. Ferrant et. al., *Synchronous Ethernet and IEEE 1588 in Telecomes*, Wiley, Hoboken, NJ, June 2013.

[Fra11]   A. Frank, *About Time: Cosmology and Culture at the Twilight of the Big Bang*, Free Press, New York, 2011.

[Fri95]   E. G. Friedman, *Clock Distribution Networks in VLSI Circuits and Systems*, IEEE Press, May 1995.

[Gar05]   F. M. Gardner, *Phaselock Techniques*, 3rd edn., Wiley-Interscience, Hoboken, NJ, July 2005.

[Ger67]   E. A. Gerber and R. A. Sykes, "Quartz frequency standards," *Proc. IEEE*, vol. 55, no. 6, pp. 783–791, 1967.

[Jon00]   T. Jones, *Splitting the Second: The Story of Atomic Time*, Institute of Physics, 2000.

[Kil64]   J. Kilby, "Miniaturized electronic circuits," U.S. patent 3138743, 1964.

[NIS09]   "NIST-F1 Cesium Fountain Atomic Clock: The primary time and frequency standard for the United States," available: http://www.nist.gov/pml/div688/grp50/primary-frequency-standards.cfm

[Noy61]   R. N. Noyce, "Semiconductor device-and-lead structure," U.S. patent 2981877, 1961.

[Oma09]   "OMAP3530/25 applications processor," SPRS507F, Rev. F, Texas Instruments Inc., Oct. 2009, available: http://www.ti.com/lit/ds/symlink/omap3530.pdf.

[Raz03]   B. Razavi, *Phase-Locking in High-Performance Systems: From Devices to Architectures*, Wiley-IEEE Press, Hoboken, NJ, Feb. 2003.

[Vit88]   E. Vittoz, M. Degrauwe, and S. Bitz, "High-performance crystal oscillator circuits: Theory and application," *IEEE J. Solid-State Circuits*, vol. 23, no. 3, pp. 774–783, June 1988.

[Wal95]   F. L. Walls and J. R. Vig, "Fundamental limits on the frequency stabilities of crystal oscillators," *IEEE Trans. Ultrason. Ferroelectr. Freq. Control*, vol. 42, no. 4, pp. 576–589, 1995.

2

EVERYTHING ABOUT THE CLOCK

As the timekeeper in the electrical world, a clock is both simple and complex. It is simple because a clock is just a train of electrical pulses that requires only one distinguished characteristic for identification: its frequency. It is complex since there are many issues around the clock: clock generation, clock distribution (logical and physical), clock usage (from system and electrical perspectives), clock quality, and clock network power consumption.

2.1 CLOCK GENERATION

A clock signal is uniquely identifiable by its frequency. Thus, clock signal generation is commonly referred to as frequency synthesis. In general, all frequency synthesis techniques can be classified into two approaches, direct and indirect, as illustrated in Figure 2.1. Direct analog frequency synthesis generates other frequencies from two or more source frequencies directly by using nonlinear electrical devices (such as a diode) which can produce frequency that is not presented at its sources [Gil82, Rok98]. Frequency filtering is often used in this method to selectively pass (or block) frequencies from one stage to the next. Direct digital frequency synthesis utilizes a fixed-high-frequency clock as an external reference to produce other lower frequencies through direct sinusoidal waveform construction. The digital data for composing the sinusoidal waveform are read from storage in every reference clock cycle and fed into a DAC (digital-to-analog converter). The resulting waveform is the

From Frequency to Time-Average-Frequency: A Paradigm Shift in the Design of Electronic Systems, First Edition. Liming Xiu.
© 2015 The Institute of Electrical and Electronics Engineers, Inc. Published 2015 by John Wiley & Sons, Inc.

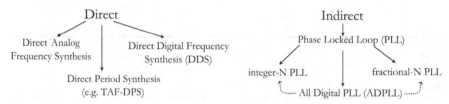

FIGURE 2.1   Clock generation: methods of frequency synthesis.

clock signal with the desired frequency [Kro98, Gol99]. Indirect frequency synthesis refers to PLL-based techniques. It is called indirect because it uses feedback and the output is not connected to the reference in a direct way (Figure 1.6). In modern VLSI system design, the PLL is the mainstream on-chip clock generator due to its ease of on-chip integration and low implementation cost. In recent years, a new approach of direct period synthesis has emerged as an enhancement to PLL-based techniques [Mai00, Cal02, Xiu02, Sot10a, Sot10b, Sot10c, Tal12, Tal13]. Starting from a base time unit, a direct period synthesizer creates a clock pulse train by individually controlling each pulse's length-in-time through counting the number of the base units in it [Xiu12]. This technique is termed TAF-DPS (Time-Average-Frequency direct period synthesis). The aim of this approach is to solve the two long-lasting problems in this field: arbitrary frequency generation and instantaneous frequency switching. This is the implementing technology for the Time-Average-Frequency based clock generator. It is the circuit-level enabler for system-level innovation. More discussion will be carried out in later chapters through examples.

2.2   CLOCK DISTRIBUTION: FUNCTIONAL

Modern chips are designed using a cell-based methodology. The chip is made of two types of cells: logic cells (also called combinational cells) and sequential cells. Logic cells are used to manipulate information. They are not controlled by a clock. The actions associated with them happen continuously (in a nonstop fashion): The input is checked all the time and the output is always updated. A sequential cell is the one that stores information. It is controlled by a clock signal: Information update only occurs at the clock edge (the moment of the level-crossing-a-threshold in Figure 1.5). Therefore, the clock signal needs to be delivered to every sequential cell [Fri95]. In this clock delivery task, there are three terms used to identify the physical and logic objects involved: *clock source*, *clock sink* (or clock leaf), and *clock domain*. Clock source refers to the circuit element that generates clock pulses. Typically it is the on-chip PLL or other types of on-chip frequency synthesizers. Clock sink is the circuit element that receives the clock pulses. It is the sequential cell (flip-flop, latch, memory, etc). A clock domain is a group of clock sinks (and their associated logic cells) wherein all the cells are controlled by the same clock signal. Each clock domain has a unique operating frequency. Due to the great complexity of a large chip

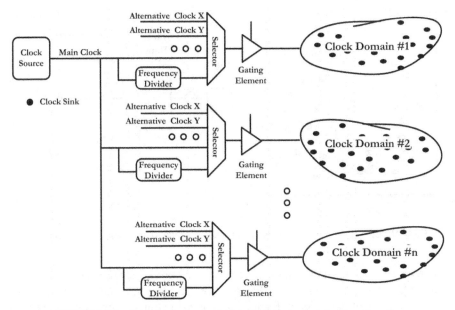

FIGURE 2.2   Distributing a clock signal to all clock sinks, functionally.

performing many functions, cells in a large chip are usually divided into different clock domains. To make the case more complicated, a particular cell could belong to different clock domains in different operating modes (such as function mode vs. test mode). As a result, a clock distribution plan is surely required to determine *which clock sink receives what clock and when*. Figure 2.2 is such a clock distribution plan while Figure 1.11 is an example of one. There are three circuit components needed in a clock distribution plan: frequency divider, clock selector, and clock gating element. The divider is used to generate a new frequency from a higher frequency source. The selector is for selectively passing a clock of a desired frequency. It is especially useful for accommodating various operating modes. The clock gating element is designed to shut down an entire clock domain at a desired time for intended purposes, such as reducing power consumption. For a large chip, the clock distribution plan can be very complex (see the example in Figure 1.11).

2.3   CLOCK DISTRIBUTION: PHYSICAL

The other aspect of clock distribution is physical related. Inside a chip, clock sources and sinks are physically located at their assigned locations, as illustrated on the left-hand side in Figure 2.3. They need to be connected through metal lines so that electrical pulses can be passed through. When a clock signal travels through these metal lines, its strength is gradually weakened due to the energy dissipated on parasitic resistance. Buffer cells are therefore used to compensate for this energy loss and to

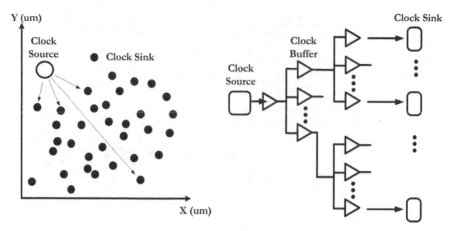

FIGURE 2.3   Distributing a clock signal: physical view (left); logical view (right).

maintain the clock signal's voltage level at an appropriate value along the delivery paths. These buffer cells are typically configured in a tree structure (clock tree), as shown on the right-hand side in Figure 2.3. In constructing this tree, two important requirements need to be fulfilled and they are all timing related. First, the time taken to deliver the clock pulses from source to sinks (termed clock tree insertion delay) has to be as short as possible. Second, a clock pulse must reach all the sinks at the same time. The maximum difference is quantified by clock skew. These requirements play a critical role in determining the quality of the chip since they have direct impact on *setup* and *hold* margins [Xiu07, Xiu12]. Besides the buffer tree, the mesh and grid are also popular methods of clock distribution. Skew and delay minimization are their design targets.

2.4   CLOCK USAGE: SYSTEM PERSPECTIVE

From a system perspective, the role of the clock is as illustrated in Figure 2.4. In a baseband signal processing system, the clock is used for information slicing (analog-to-digital process) and deslicing (digital-to-analog process). In RF applications, the demodulator and modulator associated with the RF receiver and transmitter, respectively, are also controlled by the clock. For the digital signal processing unit in the center, the clock is also the driver.

2.5   CLOCK USAGE: ELECTRICAL PERSPECTIVE

The clock is used in an electronic system to control the sequential cells. Its purpose is to ensure that, in the electronic system, information (or data) is processed and transferred in a stage-by-stage fashion. A sequential cell has at least three ports: data input port, data output port, and clock port. The typical D-type flip-flop (DFF) shown

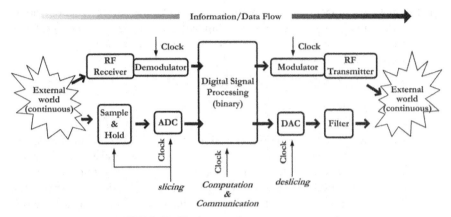

FIGURE 2.4   Clock usage from system perspective.

on the left in Figure 2.5 is a commonly used sequential cell. Its three ports are labeled D (input), Q (output), and CLK (clock). For a sequential cell to function correctly, a user must ensure that the input data and the clock signal do not change simultaneously. Otherwise, the electrical level of the sequential cell's output is nondeterministic within a finite time window. Within this window, depending on the circuit condition, it could be interpreted as either logic 1 or logic 0 and thus be unpredictable (and unusable). This is the problem of metastability [Alt09]. Eventually, given enough time, the output's electrical level will level out as either 0 or 1, but the user (circuit designer) has no control over how long this will take.

The right-hand side of Figure 2.5 presents real data regarding metastability in a sequential cell. As shown, the CLK and data signals are intentionally changed at the same time. As a result, the output Q is unpredictable. Sometimes, it is at logic level low. At other times, it is at level high. For this reason, two types of timing checks are established around the clock signal, *setup* and *hold*, as shown in the middle of Figure 2.5. Their purpose is to prevent this scenario from happening during chip operation. These checks require that, within a finite time window when the clock

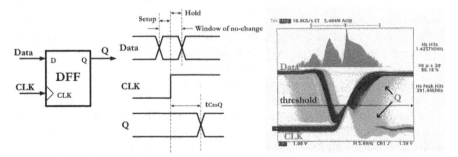

FIGURE 2.5   Typical sequential cell: DFF (left), setup and hold checks (middle), and DFF output at metastable state (right, obtained from Tektronix website, www.tek.com).

signal is making a transition, the input data cannot change. For reliable operation electricalwise, all the sequential cells in a chip have to pass these two checks. Only by performing these checks, can information be transferred from one stage to another stage in a controlled and safe way.

2.6   CLOCK SIGNAL QUALITY

In an abstract view, a clock signal is depicted in Figure 1.5 where all the timespans between any two adjacent moments in time $t_1, t_2, t_3, \ldots$ are equal and its value is called the clock period T ($\equiv 1/f$, where f is the clock frequency). During implementation, however, this is seldom the case. These voltage-crossing-a-threshold points are not always at their designed positions. This clock edge fluctuation from its ideal location, called *jitter*, is an important parameter for gauging the quality of a clock signal. Due to its importance and complexity, jitter has been quantified from a variety of aspects for various purposes: period jitter, absolute jitter, cycle-to-cycle jitter, long-term jitter, accumulated jitter, root-mean-square (RMS) jitter, peak-to-peak (P2P) jitter, periodic jitter, total jitter, etc. [Tek07]. Jitter can also be classified by the nature of the fluctuation: random or deterministic.

Jitter is a time domain parameter. Clock edge uncertainty can be studied from the frequency domain as well. A clock signal can be modeled by $v(t) = A \cos[2\pi f t + \varphi(t)]$, where f is the clock frequency and $\varphi(t)$ is the phase. The fluctuation of the voltage-crossing-a-threshold points can be viewed as noise added to the phase $\varphi(t)$, resulting in another term, *phase noise*. This term is typically used by RF engineers and serial communication link designers while the term jitter is mainly used by digital engineers. The two terms are related [Abi06, Dem06, Ric01] and are resulted from looking at the same physical phenomenon from different viewpoints.

In digital system design, the most often used jitter terms are period jitter and cycle-to-cycle period jitter. P2P period jitter is commonly used to specify a clock that will drive a synchronous system. Due to the fact that random jitter is unbounded, P2P period jitter must be qualified with a BER (bit error rate) value, for example, 10^{-12}. RMS period jitter is the standard deviation calculated from a group of period samples. According to the JEDEC (Joint Electron Device Engineering Council) standard, the required sample size is 10,000. P2P period jitter and RMS period jitter are related though a multiplier at each BER value. For example, for a BER value of 10^{-4}, the multiplier is 7.43 (P2P = 7.43 × RMS). For a BER value of 10^{-12}, the multiplier is about 14. Cycle-to-cycle period jitter is the measurement of the period difference between two adjacent cycles. It is usually used to specify a clock of varying frequency when it is not possible to use the P2P period jitter specification, as in the case of spread spectrum clock generation.

In terms of specifying clock quality, a serial communication link is much more complicated than an synchronous system. In this application, the TX (transmitter) and RX (receiver) both have embedded clock circuits. The TX and RX interact in such a way that, in terms of jitter frequency content, some parts of jitter play a more important role than other parts. Thus, jitter needs to be clearly revealed and/or specified. In this scenario, a frequency domain description is the appropriate means

TABLE 2.1   RMS Phase Jitter Integration MASK

Standard	Integration MASK
Fiber channel	637 KHz $\leftrightarrow$ 10 MHz
10-G Ethernet XAUI	1.875 MHz $\leftrightarrow$ 20 MHz
SONET OC-48	12 KHz $\leftrightarrow$ 20 MHz
SONET OC-192	20 KHz $\leftrightarrow$ 80 MHz
SATA/SAS	900 KHz $\leftrightarrow$ 7.5 MHz

of breaking down jitter components. The phase noise plot is typically generated by a spectrum analyzer that captures the spectral content of a clock signal. It is useful for studying the frequency characteristics of the jitter. In a clock waveform, most of the energy is located at the fundamental frequency (the carrier frequency). However, some portion of the energy leaks out to a range of frequencies around the carrier. The term *phase jitter* is used to specify the amount of phase noise energy contained between two offset frequencies relative to the carrier. The RMS phase jitter is the integration of the phase noise power over a specified frequency band (called the MASK). It can be specified in units of time (e.g., picoseconds and femtoseconds) or in UI (unit interval) for a particular data rate. Some application-specific filters (MASK) are listed in Table 2.1. RMS phase jitter determined by integrating phase noise power over the selected MASK is useful in specifying which part of the jitter has an influence on the system's performance and which part can be ignored. This type of jitter specification is especially useful for serial link design.

2.7   CLOCK NETWORK POWER CONSUMPTION

Among all the on-chip signals, the clock has the largest fanouts since it needs to be delivered to all sequential cells in a chip. It also has the highest switching activity because it toggles every cycle. For these reasons, the clock network typically uses the largest percentage of the total power consumption of a chip. A possible way of reducing clock power consumption is to recycle the energy used by the clock network through *LC* resonance [Drk04, Hu12]. A more conventional approach is dynamically adjusting the clock frequency based on system loading since power consumption is directly proportional to the operating frequency [Nak05, Mal06, Sal11]. A third possible way is to lower the frequency of the global clock signal. After it is delivered to the local modules, the clock signal's frequency is then boosted to support the modules' operation. These techniques will be addressed in later chapters in detail.

REFERENCES

[Abi06]   A. A. Abidi, "Phase noise and jitter in CMOS ring oscillators," *IEEE J. Solid-State Circuits*, vol. 41, no. 8, pp. 1803–1816, Aug. 2006.

[Alt09]   "Understanding Metastability in FPGAs," White Paper, WP-01082-12, Altera, July 2009. http://www.altera.com/literature/wp/wp-01082-quartus-ii-metastability.pdf

[Cal02]   D. E. Calbaza and Y. Savaria, "A direct digital periodic synthesis circuit," *IEEE J. Solid-State Circuits*, vol. 37, no. 8, pp. 1039–1045, 2002.

[Dem06]   A. Demir, "Computing timing jitter from phase noise spectra for oscillators and phase-locked loops with white and 1/f noise," *IEEE Trans. Circuits Syst. I*, vol. 53, no. 9, pp. 1869–1884, Sept. 2006.

[Drk04]   A. J. Drake, K. J. Nowka, T. Y. Nguyen, and J. L. Burns, "Resonant clocking using distributed parasitic capacitance," *IEEE J. Solid-State Circuits*, vol. 39, no. 9, pp. 1520–1528, Sept. 2004.

[Fri95]   E. G. Friedman, *Clock Distribution Networks in VLSI Circuits and Systems*, IEEE Press, May 1995.

[Gil82]   B. Gilbert, "A monolithic microsystem for analog synthesis of trigonometric functions and their inverses," *IEEE J. Solid-State Circuits*, vol. 17, no. 6, pp. 1179–1191, Dec. 1982.

[Gol99]   B. G. Goldberg, *Digital Frequency Synthesis Demystified*, LLH Technology Publishing, 1999.

[Hu12]   X. Hu and M. R. Guthaus "Distributed LC resonant clock grid synthesis," *IEEE Trans. Circuits Syst. I*, vol. 59, no. 11, pp. 2749–2760, Nov. 2012.

[Kro98]   V. F. Kroupa, *Direct Digital Frequency Synthesis*, IEEE Press, 1998.

[Mai00]   H. Mair, and L. Xiu, "An architecture of high-performance frequency and phase synthesis," *IEEE J. Solid-State Circuits*, vol. 35, pp. 835–846, June 2000.

[Mal06]   A. Mallik et al., "User-driven frequency scaling," *Comput. Architect. Lett.*, vol. 5, no. 2, p. 16, 2006.

[Nak05]   M. Nakai et al., "Dynamic voltage and frequency management for a low-power embedded microprocessor," *IEEE J. Solid-State Circuits*, vol. 40, no. 1, pp. 28–35, Jan. 2005.

[Ric01]   R. Poore, "Overview on phase noise and jitter," Agilent Technologies, 2001, available: http://cp.literature.agilent.com/litweb/pdf/5990-3108EN.pdf.

[Rok98]   A. Rokita, "Direct analog synthesis modules for an X-band frequency source," paper presented at MIKON '98, 12th International Conference on, Microwaves and Radar, 1998, May 20–22, 1998, vol. 1, pp. 63–68.

[Sal11]   M. E. Salehi et al., "Dynamic voltage and frequency scheduling for embedded processors considering power/performance tradeoffs," *IEEE Trans. VLSI*, vol. 19, no. 10, pp. 1931–1935, Oct. 2011.

[Sot10a]   P. Sotiriadis, "Theory of flying-adder frequency synthesizers part I: Modeling, signals periods and output average frequency," *IEEE Trans. Circuits Syst. I*, vol. 57, no. 8, pp. 1935–1948, Aug. 2010.

[Sot10b]   P. Sotiriadis, "Theory of flying-adder frequency synthesizers part II: Time and frequency domain properties of the output signal," *IEEE Trans. Circuits Syst. I*, vol. 57, no. 8, pp. 1949–1963, Aug. 2010.

[Sot10c]   P. Sotiriadis, "Exact spectrum and time-domain output of flying-adder frequency synthesizers," *IEEE Trans. Ultrason. Ferromagnet. Freq. Control*, vol. 57, no. 9, pp. 1926–1935, Sept. 2010.

[Tal12]   S. Talwalkar, "Quantization error spectra structure of a DTC synthesizer via the DFT axis scaling property," *IEEE Trans. Circuits Syst. I*, vol. 59, no. 6, pp. 1242–1250, June 2012.

[Tal13]   S. Talwalkar, "Digital-to-time synthesizers: Separating delay line error spurs and quantization error spurs," *IEEE Trans. Circuits Syst. I*, vol. 60, no. 10, pp. 2597–2605, June 2013.

[Tek07]   Tektronix, "Understanding and characterizing timing jitter," Application Note, 55W-16146-3, Tektronix, 2007.

[Xiu02]   L. Xiu and Z. You, "A Flying-Adder architecture of frequency and phase synthesis with scalability," *IEEE Trans. VLSI*, vol. 10, pp. 637–649, Oct. 2002.

[Xiu07]   L. Xiu, *VLSI Circuit Design Methodology Demystified: A Conceptual Taxonomy*, Wiley-IEEE Press, Hoboken, NJ, Nov. 2007.

[Xiu12]   L. Xiu, *Nanometer Frequency Synthesis beyond Phase Locked Loop*, Wiley-IEEE Press, Hoboken, NJ, Aug. 2012.

3

A DIFFERENT WAY OF CONSTRUCTING A CLOCK SIGNAL: TIME-AVERAGE-FREQUENCY

3.1 MOTIVATION

This chapter will discuss the creation of a sense-of-time in the electronic world. The common approach used today for establishing a time scale on-chip is by using a PLL, which is a beautiful blend of analog and digital circuit technologies. It is one of the foundational circuit components in today's semiconductor IP portfolio. However, as mentioned in previous discussion, there are still two problems with today's PLL-based clock generators: The available frequencies are limited and the switching between frequencies is slow. The source of the problem originates from the fact that electrical circuits are made to handle magnitude (or level), not time. In a circuit, information is represented by the medium of the electron. It is created based on the magnitude of electron flow using proportional (analog) or binary (digital) relationships. Time is created indirectly through a voltage-level-crossing-a-predetermined-threshold. Therefore, the task of building a timekeeper in an electronic system is inherently difficult because the problem is unique and the uniqueness lies in the fact that a timekeeper circuit is not just for information processing; it relates two basic properties of the universe: *time* and *force* [Bar02].

In the practice of circuit implementation, another fact has made the task of creating the sense-of-time in the electronic world even more challenging: *From the first day that the clock signal is introduced into the field of VLSI design, it is assumed that all the pulses in a particular clock pulse train have to be equal in length* (the implementation imperfection, such as clock jitter, is not considered in this statement). This supposition

From Frequency to Time-Average-Frequency: A Paradigm Shift in the Design of Electronic Systems, First Edition. Liming Xiu.
© 2015 The Institute of Electrical and Electronics Engineers, Inc. Published 2015 by John Wiley & Sons, Inc.

has limited our options in the creation of timekeeping circuits. This problem, however, is not inherited from the nature as the first one. It is caused by our perspective on clock. It is created by us. In our social life, we use a wall clock and a wrist watch to tell time and organize events. The time scales used are the year, day, hour, and minute. The base unit is the second. The time scales of year, day, hour, and minute are created from counting certain predefined values of the base unit (the second). Social events are indexed by numbers, which are counted from a starting point using a chosen time scale. The starting point agreed upon today by the world is the year that Jesus Christ was born (the birth of Christ). For example the event of the 20th FIFA World Cup takes place in A.D. 2014, which is 2014 years from the year that Christ was born. The time scale used here is the *year*. If more temporal detail about this event is demanded, it can be further specified as June 12th to July 13th. In this case, the time scale of *day* is used. If even more detail is required, the time scales of the hour, minute, and even second can be used. In this timekeeping mechanism, we are used to the fact that every second is exactly the same as the next second and the previous second. This appreciation of time is sufficient for our daily life. The reason is that a second is a very small time granularity compared to the time span of most daily events in our lives. Therefore we do not care about its fine structure at this granularity level; making every second exactly the same is the most cost effective way of building clockwork.

Since the second is the only time unit adopted by the world that is precisely defined and its accuracy properly maintained, the designer of an electronic circuit must use it as well to establish the reference for the time flow inside the electronic circuit so that the social world and the electronic world can be related. The two worlds, however, use the second in different ways. The difference is caused by the fact that, although a second is a very fine unit for social events, it is an extremely large time span for events happening in the electronic world. In other words, in the electronic world, the pace of life is much faster compared to that of the social world. Table 3.1 lists some facts to illustrate this point. Hence, due to the different application scenarios, the second is used differently. In the social world, the second is used as a base for counting when events happen. In the electronic world, the second is used as *a fixed frame* to count the number of a selected base unit. This base unit is the length (measured in time) of an electrical pulse (the clock period; refer to Figure 1.5).

The mechanisms of using the second are also different in these two worlds. In the social world, we care what event happens at what time. We identify events by labeling them using a starting point number that steps forward at each count. In other words, we number index each event explicitly. On the other hand, in the electronic world, we focus our attention more on the relative temporal order of events. The absolute

TABLE 3.1   Second Scale in Timing Social and Electronic Events

Social world	year/s = 3.15×10^7	day/s = 86,400	h/s = 3600	min/s = 60
Electronic world	1-MHz clock	100-MHz clock	1-GHz clock	10-GHz clock
	base^a/s = 10^{-6}	base/s = 10^{-8}	base/s = 10^{-9}	base/s = 10^{-10}

^{a}Base is the clock pulse's length-in-time, or the clock period.

indexing of events is not as important. From a chip's point of view, all the devices inside the chip make the whole world. Thus, the starting point of life can be easily set and reset (e.g., the chip-power-up or reset-button-being-pushed moment). As a means for coordinating tasks, identifying the temporal order of events is more frequently used than labeling each event. The key differences between the two mechanisms are as follows:

- The temporal order of events is established by the forward flow of seconds in the social world. The temporal order of events is established by the forward flow of an electrical pulse train in the electronic world.
- The second is used explicitly in the timekeeping mechanism of the social world. It is used implicitly in the electronic world through an electrical pulse train.
- Events are number indexed in the social world with a fixed starting point. The starting point for identifying events is frequently reset in the electronic world where the focus is on the relative temporal order of the events.

The bridge that links the electrical pulse train and the second is the clock frequency. The pace of an electronic system's operation is controlled by the electrical pulse train. The basic counting unit is the individual pulse, not the second as in the social world. Pulses are used to differentiate events. This is all we want from the clock pulse train. Therefore, a fundamental question can be asked: Do all the pulses in a clock pulse train have to be equal in length? This question is equivalent to asking: *What does clock frequency really mean?* In 2008 a novel concept, time-average-frequency, was introduced [Xiu08a]. It removes the constraint that all pulses (or clock cycles) must be equal in length. It is based on the understanding that clock frequency is used to indicate the number of operations executed (or events happening) within the time window of 1 s. As long as the specified number of operations is completed successfully in a specified time window (such as 1 billion operations in 1 s for a 1-GHz CPU), the system does not care (or cannot tell the difference between) how each operation is carried out in detail.

In the social world, we use time in the sense of *what happens at when*. In the electronic world, we often compare systems by *how many things happened (can happen) within a fixed time window*. In both cases, the second is used as the reference: one as a base unit and the other as a time window. It is true that, as a base unit, any second has exactly the same length-in-time as any other second. But, this fixed-length feature is just a convenience, not an absolute necessity. To determine the temporal order of events, which is the essence of timekeeping, both fixed-length and variable-length base time units can be used. When the second is used as a time frame to count another base unit, there is no fundamental reason that that base unit has to be fixed in its length-in-time. The selection of which scheme to use should be decided by the implementation cost. Table 3.2 highlights the role of the second in identifying events in the social and electronic worlds. As shown, due to the nature of electrical operation, many time scales are needed. Furthermore, a change of time scale occurs frequently in the action. This is the motivation for demanding a flexible clock generator capable

TABLE 3.2 Arranging Events in Temporal Order in Social and Electronic Worlds

	Time Scale	Starting Point	Role of Second	Flexibility of Time Scale
Social world	Year, day, hour, minute	Birth of Christ (AD)	As a base for being counted	Year/day/hour/ minute is sufficient; events arranged using fixed scale
Electronic world	Clock period (pulse's length)	Every moment of power-up or reset	As a frame for counting other base units	Countless scales needed; change of scale frequently required

of generating many time scales. The variable-length base unit can make the task of constructing such flexible clock generator possible.

Time-average-frequency is a new type of clock structure that serves the same purpose as the conventional clock but with more flexibility. This breakthrough in the clock frequency concept is crucial. Time-average-frequency provides chip designers with the means of making flexible on-chip clock sources in a cost-effective way.

This concept and its implementation technology provide a link between circuit and system: a circuit-level enabler for system-level innovation.

3.2 CLOCK IS TRIGGER AND GATEKEEPER: ESSENCE OF STAGE-BY-STAGE OPERATION

In modern chip design the mainstream design method of establishing temporal order among events is through stage-by-stage operation: the synchronous design method. In this approach, a chip in a VLSI system is structurally made of combinational and sequential cells. Data are processed by combinational cells. Data are created and moved between stages whose boundaries are established by sequential cells. Each sequential cell is a storage room whose door is controlled by a clock signal. Most of the time, the door is closed. Only when the clock signal makes a low-to-high or high-to-low transition (clock edge) does the door open temporarily; the old data move out and new data move in. In this regard, the clock signal is the trigger for logic operation and the gatekeeper for data flow, as illustrated on the left in Figure 3.1.

In this stage-by-stage operation, there are three timing parameters involved. The first one represents the speed of the circuit located between stages, t_{cir}, and it is stage dependent. The second one is related to the speed of the open-then-close action of the sequential circuit (the door speed of the storage room), t_{oc}. It is cell-type (room-type) dependent. The last one is the clock speed T. It is the period of all the clock cycles, as shown on the right-hand side of Figure 3.1. The stage-by-stage operations are accomplished by launch-and-receive actions, which have two aspects: physical and

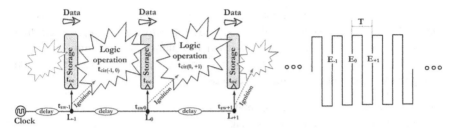

FIGURE 3.1   Clock signal is the trigger and gatekeeper (left) and the clock waveform for establishing temporal order (right).

temporal. As depicted on the left of Figure 3.1, there are three storages locations: L_{-1}, L_0, and L_{+1}. When L_0 is considered as a receiver, L_{-1} is the launcher. In operation, a storage cell functions as both a launcher and a receiver (e.g., L_0 is the launcher when L_{+1} is concerned). When investigated in a temporal relationship, the launch-and-receive pair is made of two clock edges. For example, in the right-hand drawing, E_{-1} is the launching edge and E_0 is the receiving edge. Similarly, in the temporal perspective, a clock edge works as both a launcher and a receiver.

The essence of stage-by-stage operation is (1) to ensure that data arrive at the destination on time and (2) to prevent data pass-through. If one stands at location L_0 at moment E_{-1}, within one clock cycle (e.g., the time frame established by the two edges E_{-1} and E_0), data launching from L_{-1} is allowed to travel only to L_0, not to L_{+1} or beyond. This statement can be translated into two requirements: (1) Data cannot travel too slow and (2) data cannot travel too fast. Assume data are launched from L_{-1} at E_{-1}. It will not be latched into storage at L_0 if its traveling time is longer than T ($t_{cir(-1,0)} > T$) since the door at L_0 is already closed when it reaches the destination (i.e., the moment of E_0 is passed). This scenario is prevented from happening by the *setup* check performed during logic synthesis. This check involves two locations (L_{-1} and L_0) and two moments (E_{-1} and E_0).

On the other hand, there is the danger of pass-through if data travel too fast. If one stands at location L_0 at moment E_0 (not E_{-1} as in the previous case), he or she sees data coming from L_{-1} and approaching L_0. If the incoming data travel fast enough ($t_{cir(-1,0)} < t_{oc}$), the storage at L_0 will not have enough time to close the door (and thus keep the data). This operation is also faulty and it is prevented by the *hold* check. This check involves two locations (L_{-1} and L_0) and ONE moment (E_0).

In real circuit operation, when a clock signal's physical distribution is included in the consideration, there is one more fact that further complicates the situation. When a clock signal travels from its source to L_{-1}, L_0, and L_{+1}, the arrival times are different due to the different delivery paths. This is called clock skew. Its impact is that it changes the clock period T that each storage cell sees. Its influence on the SoC time closure must be considered as well. More discussion on this issue can be found in Sections 1.3 and 4.22 of [Xiu12a]. When both the setup and hold checks are satisfied, the metastability condition is avoided and the data stored in each sequential cell will be the intended data.

In summary, the task of stage-by-stage operation, which is the foundation of modern chip design, can only be understood by establishing a clear picture according to the points given below. This picture is crucial for appreciating the mechanism of using a clock pulse train to establish temporal order in the circuit. It is a crucial step leading to the time-average-frequency concept.

- Three circuit related timing parameters: t_{cir}, t_{oc}, T
- One physical-related timing parameter: t_{sw}
- A physical launch-and-receive pair (e.g., L_{-1} and L_0)
- A temporal launch-and-receive pair (e.g., E_{-1} and E_0)
- Setup timing check (preventing slow circuit from being created): involves two locations and two moments; checks circuit speed against clock period.
- Hold timing check (preventing fast circuit from being created): involves two locations and one moment; checks circuit speed against the speed of the open-then-close action.

3.3 TIME-AVERAGE-FREQUENCY: BRIEF REVIEW

By examining Figure 3.1, a question can be naturally asked: Do all the cycles in a clock pulse train have to be equal in length-in-time? In other words, can T take different values for different cycles? From the discussion in Section 3.2, it can be seen that the stage-by-stage operation can still be successfully carried out even in the scenario that clock cycles have different lengths in time. During the circuit's logic synthesis (the step that turns a circuit's RTL (Register Transfer Level) description into logic gates), the setup check can be satisfied as long as the smallest value of T is used as the setup constraint. The hold check is not affected by the variation in the value of T since the hold check uses only one time moment (it has nothing to do with T). This understanding is established from the perspective of the clock signal user. It can lead to a new direction in clock generator circuit design when this understanding is passed onto its designer.

Based on this understanding, the time-average-frequency concept was established in 2008 [Xiu08a]. Using this approach, instead of one type of pulse, two or more types of pulses can potentially be used in a clock pulse train. In Figure 3.2, where waveforms of both conventional frequency and time-average-frequency clocks are

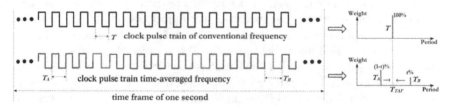

FIGURE 3.2 Waveforms of conventional frequency (top) and time-average-frequency (bottom).

shown, there is only one type of pulse T in the conventional frequency clock signal (upper trace). In the time-average-frequency case (bottom trace), two types of pulses, T_A and T_B, are used. The period distributions of these two types of clock signals are depicted in the drawings on the right side, where r is the weighting factor between T_A and T_B:

$$\frac{1}{f_{TAF}} = T_{TAF} = \frac{\left(\sum_{i=1}^{N_{TAF}} T_i\right)}{N_{TAF}} \quad \text{where} \quad T_i \text{ can be either type } T_A \text{ or type } T_B \quad (3.1)$$

$$\frac{1}{f_{FD}} = T_{FD} = \sum_{i=1}^{N_{TAF}} T_i \quad \text{where} \quad T_i \text{ can be either type } T_A \text{ or type } T_B \quad (3.2)$$

The formal definition of time-average-frequency was introduced in [Xiu08a]. Mathematically, the time-average-frequency $f_{TAF} = 1/T_{TAF}$ is expressed in (3.1), where N_{TAF} is the minimum number of cycles that makes the clock waveform repeat. Fundamental frequency $f_{FD} = 1/T_{FD}$ is defined in (3.2) and $T_{FD} = N_{TAF}T_{TAF}$. The time-average-frequency is different from the often loosely used term "average frequency" in that $f_{TAF} = 1/T_{TAF}$ is only defined in N_{TAF} while average frequency can be defined in any value of N. In the frequency domain, the spectrum of f_{TAF} can be precisely calculated using the theory developed in recent years [Gui10, Sot10a, Sot10b, Sot10c, Tal12, Tal13, Xiu08b, Xiu10, Xiu11]. For those who are new to this concept or those who still have doubts, an example is provided below to assist the reader in understanding the spirit of time-average-frequency.

Assume that a government agency plans to establish a housing assistance program to help its people. The program is funded to build about one million houses from location A (e.g., Houston, TX) to location B (Dallas, TX). Ideally, all the houses are supposed to be built on lots of equal size. The distance between the two locations is given as L. Therefore, the size of each lot is $L/1,000,000$ (assume further that the widths of all the lots are as fixed value). Under the guideline that all the lots must be the same size, exactly one million families can be accommodated if the builder chooses $L/1,000,000$ as the standard lot size. However, the agent and the builder have to take into consideration the small error that could occur on the estimation of the population. For example, if the number of applicants is not exactly 1,000,000 but 999,999, the builder must use a lot size of $L/999,999$ to fulfill the size requirement. Even worse, this number could change considerably before the builder can finalize the plan on lot size or even after construction is started.

A solution to this problem is for the builder to premanufacture two types of lots: U1 ($=L/1,000,000$) and U2 $= (1+x\%)$U1, where x is a small number which could be either positive or negative. By using these lots in an interleaved fashion, the number of lots that can fit into this L is calculated as $L = a\cdot$U1 $+ b\cdot$U2, where both a and b are integers. As an example, assume x is chosen as 5 (lot U2 is 5% larger than that of U1). Table 3.3 lists the plans to accommodate the variation on the number of families that apply for this program. By adopting this strategy, the builder can start the development process without knowing the exact number of applicants. If needed, the overall landscape can be adjusted at a later stage.

TABLE 3.3 Using Two Types of Lots to Accommodate a Variety of Scenarios

Plan	U1	U2	$a \rightarrow$ # of lot U1	$b \rightarrow$ # of lot U2	Total # of lots (families)	$a \cdot$U1 + $b \cdot$U2
1	L/1,000,000	1.05·U1	1,000,000	0	1,000,000	L
2	L/1,000,000	1.05·U1	999,979	20	999,999	L
3	L/1,000,000	1.05·U1	999,958	40	999,998	L
4	L/1,000,000	1.05·U1	999,937	60	999,997	L
5	L/1,000,000	1.05·U1	999,916	80	999,996	L
6	L/1,000,000	1.05·U1	999,895	100	999,995	L
...						

In the spirit of fairness, the criterion that "all the lots be the same size" is ideal, but it is difficult to implement in many situations. The latter approach improves the system's flexibility at the expense of slightly degraded fairness, but the main goal of helping families can still be well served. In the electrical world, the fixed distance L is the lapse of time of 1 s, a constant based on fundamental properties of nature [All97]. The lot size is the pulse length in a clock pulse train. The number of lots is the number of pulses within 1 s (i.e., the clock frequency). The roles of the government agent and the builder are equivalent to that of the timing circuit designer. A particular family served is the logic/arithmetic operation that must be carried out in a clock cycle. Table 3.3 shows that, virtually, any number of families can be accommodated by this plan (within a certain range). The spirit of time-average-frequency is to use two (or more) types of pulses to construct the clock pulse train and then to fulfill the need of complex application scenarios. Similarly, virtually any clock frequency can be produced by adopting time-average-frequency.

An important point worth mentioning is that this mechanism of generating a large number of frequencies is not achieved simply at the circuit level (i.e., modifying the lot size). *It is accomplished at a higher level of theoretical thinking: reinvestigation of the clock frequency concept.*

3.4 CIRCUIT ARCHITECTURE OF TIME-AVERAGE-FREQUENCY DIRECT PERIOD SYNTHESIS

A circuit structure that implements the ideas expressed in (3.1) is the direct period synthesizer, as depicted in Figure 3.3. This method is termed time-average-frequency direct period synthesis (TAF-DPS) since each pulse length is directly synthesized, and thus its length can vary from cycle to cycle. In this method, a base time unit Δ (e.g., 25 ps) is created first. From this Δ, two (or more) types of cycles, $T_A = I\Delta$ and $T_B = (I+1)\Delta$, are formed, where I is an user-selected integer. The circuit uses T_A and T_B alternatively to make the pulse train. The time-average-frequency is generated as expressed in (3.3), where r is the weighting factor mentioned above. It is a fraction that represents the possibility of occurrence of T_A (and also T_B). In a real circuit implementation, r can be expressed as $r = p/q$, p and q are integers, and $p < q$,

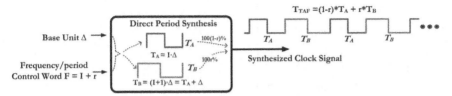

FIGURE 3.3   Circuit architecture of time-average-frequency direct period synthesis.

$\gcd(p, q) = 1$. Thus, $f_{\text{TAF}} = 1/T_{\text{TAF}}$ can also be expressed as in (3.4). It is clear that q is the N_{TAF} in (3.1). The variable F is used by TAF-DPS users as the frequency control word (more precisely, period control word) and it has two parts: the integer portion I and the fractional portion $r = p/q$. Figure 3.4 shows the building blocks of a time-average-frequency direct period synthesizer. The base time unit Δ can be generated from a multi-phase PLL, a delay chain in a delay locked loop (DLL), or a divider chain having multiple outputs. The synthesizer is made of two parts of circuitries: the upper and lower paths. Each part comprises a $K{\to}1$ multiplex and is responsible for producing the clock output high-to-low (or low-to-high) transition. One such circuit, the Flying-Adder synthesizer, is explained in great details in Chapter 4 of [Xiu12a]. The interested reader is referred to that document for further reading:

$$T_{\text{TAF}} = 1/f_{\text{TAF}} = (1 - r)T_A + rT_B \tag{3.3}$$

$$T_{\text{TAF}} = (qI + p)\Delta/q = F\Delta \quad \text{where } F = I + p/q = I + r \tag{3.4}$$

From (3.4), it is clear that the TAF-DPS period transfer function (output period vs. control word) is linear. Thus, its output frequency is inversely proportional to the frequency control word (its frequency transfer function is in $1/x$ fashion). Figure 3.5 illustrates this transfer function through transistor-level simulation. The simulation result is obtained from the circuit shown symbolically in Figure 3.3. The base time unit Δ is created from a reference frequency f_r, $\Delta = 1/(Kf_r)$, where $K = 32$ and $f_r = 20$ MHz. The frequency control word can take its value from the range $[2, 2K]$, fraction included. The TAF-DPS output frequency range is $\frac{1}{2} f_r \leq f_{\text{out}} \leq (K/2) f_r$. This is translated into 10–320 MHz. The top trace shows the frequency control word. It is stepped forward from 2 to 64 ($=2K$), one step at a time, and held for 0.3 µs per step. The bottom trace shows the output waveform from the circuit. The trace in the middle is the frequency measurement of the clock waveform at the bottom. The $1/x$ trend is clearly visible, and its range is from 320 to 10 MHz as expected. In this simulation, only integer values are used in the control word. As a result, the frequency measurement shows a stepwise curve. All the frequencies between the points can be generated by using fractions in the control word.

An interesting point worth mentioning is that the TAF-DPS can conveniently function as a frequency multiplier to boost the frequency (which usually requires the use of a PLL). This feature can be very useful to many applications, as will be demonstrated in later chapters.

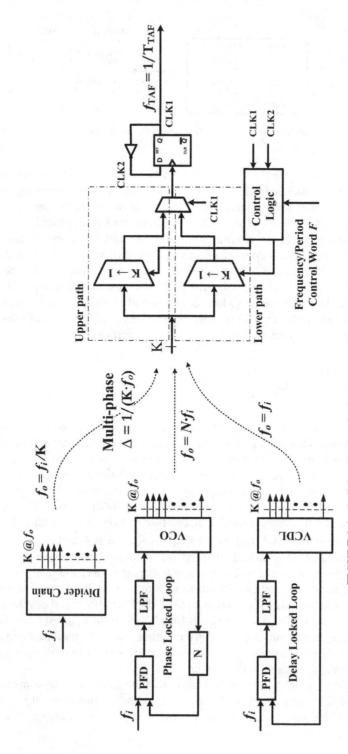

FIGURE 3.4 Building blocks of a time-average-frequency direct period synthesizer.

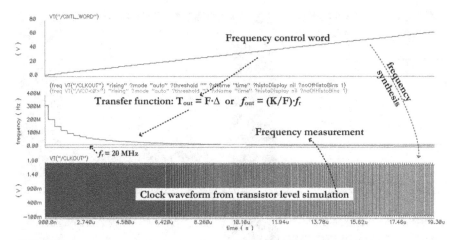

FIGURE 3.5   Frequency transfer function of TAF-DPS: control word (top), frequency measurement (middle), and clock waveform (bottom).

It is also worth pointing out that the time-average-frequency clock signal is not limited to only two types of pulses. A plurality of pulses can be used in constructing the time-average-frequency clock signal. In this generic case, the time-average-frequency is defined as in (3.5), where T_i is the length-in-time of the ith type pulse and a_i is its possibility of occurrence. Here a_i can be calculated on a segment of the time-average-frequency clock waveform. The segment is the minimum section of the clock waveform that is repeatable. In other words, the length-in-time of said segment is the fundamental period of the time-average-frequency clock signal:

$$T_{\text{TAF}} = \sum_{i=0}^{i=l} a_i T_i \qquad f_{\text{TAF}} = \frac{1}{T_{\text{TAF}}} \tag{3.5}$$

3.5   THE TWO LONG-LASTING PROBLEMS: SMALL FREQUENCY GRANULARITY AND FAST FREQUENCY SWITCHING

The purpose of introducing the TAF-DPS circuit is to address the two long-lasting problems in this field: arbitrary frequency generation (small frequency granularity) and instantaneous frequency switching. These two features can enable system-level innovations. This statement will be justified through examples in later chapters. Figure 3.6 gives an example of fine frequency resolution around a 27-MHz clock signal. The measurement plots at the top are the phase noise and the spectrum of the 27-MHz (37.03703704 ns) signal. The plots at the bottom are for a frequency of 26.9999968 MHz (37.03704146 ns), which is −3.2 Hz (−0.12 ppm) away from the 27 MHz. These

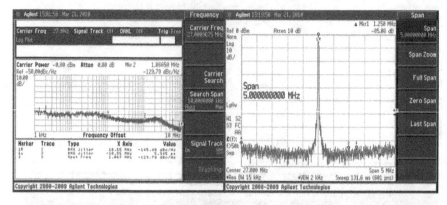

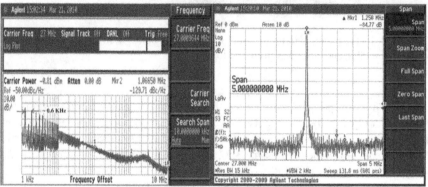

FIGURE 3.6   Fine frequency resolution example 1: phase noise and spectrum plots of 27 MHz (top) and −3.1 Hz away from 27 MHz (−0.12 ppm, bottom).

two very close frequencies are achieved by a TAF-DPS using a Δ of 192.9012346 ps and the following configurations:

27 MHz $\{I = 192, p = 0, q = 2^{17}\} \rightarrow F = I + r = 192 \rightarrow T = F\Delta = 192\Delta$
$= 37.03703704$ ns $\rightarrow f = 27$ MHz

26.9999968 MHz $\{I = 192, p = 3, q = 2^{17}\} \rightarrow F = I + r = 192 + 3/2^{17} \rightarrow$
$T = F\Delta = 192.00002289\Delta = 37.03704146$ ns $\rightarrow f = 26.9999968$ MHz

The two frequencies (clock period) are differentiated by 192Δ and 192.00002289Δ. This results in a small frequency granularity of 0.12 ppm (3.2 Hz). As shown in the measurements, the two measured frequencies are 27.0009675 and 27.0009644 MHz, respectively. There is a 3.1 Hz frequency difference and it agrees well with the calculation of 3.2 Hz. The error between the calculation and the measurement (i.e., 27 MHz vs. 27.0009675 MHz) is caused by a 36 ppm error on the 12 MHz crystal reference used in the system.

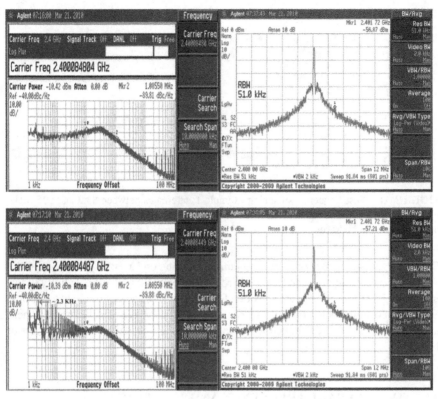

FIGURE 3.7 Fine frequency resolution example 2: phase noise and spectrum plots of 2.4 GHz (top) and −286 Hz away from 2.4 GHz (−0.12 ppm, bottom).

In the case of 26.9999968 MHz, $T_A = 192\Delta$ and $T_B = 193\Delta$ (refer to Figure 3.2). The possibility of occurrence is $r = 3/2^{17} = 0.00229\%$. Of every 131,072 cycles, there are 3 T_B cycles and 131,069 T_A cycles. As can be appreciated from this example, small frequency granularity is achieved by adjusting the ratio between T_A and T_B slightly (review the housing example in Section 3.3). If a clock signal of this frequency is made according to the conventional frequency concept (assuming it is possible), the circuit driven by this clock should be setup constrained by 37.03704146 ns (the $T = T_{TAF}$ in Figure 3.2). In the time-average-frequency clock signal, the setup constraint is 37.03703704 ns (the T_A in Figure 3.2), which is 4.42 fs tighter. This is a small price to pay for using the time-average-frequency clock in this case. Figure 3.7 is another example. It demonstrates small frequency granularity at a higher frequency of 2.4 GHz. The resolution achieved is 0.12 ppm, as shown in the measurements.

Figures 3.8 and 3.9 illustrate fast frequency switching at low frequency and at higher frequency, respectively. The speed of frequency switching is two cycles. In other words, the new clock waveform will be ready two cycles after the circuit receives the "switch" command. This can be seen on the right of Figure 3.8. The waveform in

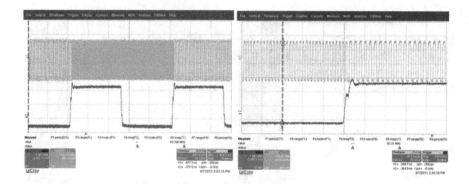

FIGURE 3.8 Fast frequency switching example 1: alternates between 83.3 and 125 MHz; plot on the right is the zoom-in version of the one on the left. In both plots, the upper trace is the output clock, the lower one is the switching control signal.

Figure 3.9 shows the case of frequency jumping among 667 MHz, 1 GHz, and 2 GHz. The clock signal stays at each frequency roughly around 20 ns. This capability of fast frequency switching is also demonstrated in Figure 3.10, where frequency hopping among three frequencies of 800 MHz, 1 GHz, and 1.33 GHz is illustrated. These measurement results are obtained from the circuit using the TAF-DPS architecture of Figure 3.3.

3.6 TIME-AVERAGE-FREQUENCY AND "JITTERY" CLOCK

Since TAF-DPS uses two (or more) types of cycles to construct a clock signal, the resulting clock signal will look jittery. Figure 3.11 illustrates this using TAF-DPS

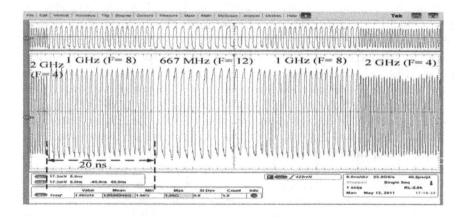

FIGURE 3.9 Fast frequency switching example 2: alternating among 667 MHz, 1 GHz, and 2 GHz.

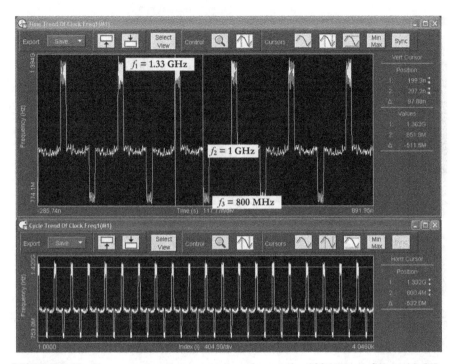

FIGURE 3.10 Frequency hopping among three frequencies 800 MHz, 1 GHz, 1.33 GHz: measured frequency vs. time trend (top); frequency vs. cycle trend (bottom, larger span).

clock waveforms captured with an oscilloscope. On the left-hand side, two waveforms of the data (top trace) and clock (bottom trace) are shown on the oscilloscope screen. The data are generated from a CDR (clock data recovery) circuit where the time-average-frequency concept is used. In other words, the data are generated from a circuit driven by a time-average-frequency clock. The oscilloscope is operated in the single-run mode. As such, it is clearly shown that two types of clock cycles existed in the clock pulse train. For example, in the circled area in the bottom, the two clock

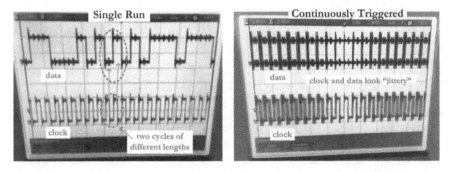

FIGURE 3.11   Snapshots of clock and data waveforms taken from Agilent MSO oscilloscope.

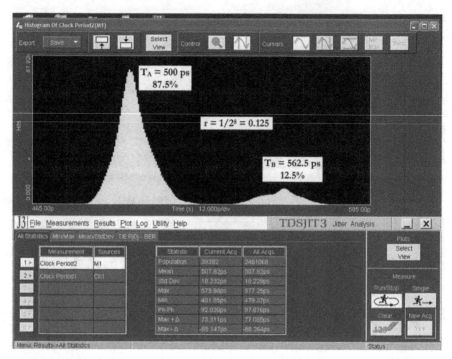

FIGURE 3.12   Measured period distribution of a time-average-frequency clock signal.

cycles' lengths are different. In the upper trace, the data associated with these two clock cycles (a 0 followed by a 1, as highlighted by the dot circle in the top) also show the fact of different lengths. On the right-hand side, the same signals are displayed but with the oscilloscope in a continuously triggered mode. Without surprise, both the data and clock look jittery. However, what important is that the circuit works correctly. In other words, the data are correctly generated by this time-average-frequency based circuit. This is confirmed by a BERT (bit error rate test) instrument used in this case. This example brings up a crucial point: *A "jittery" clock is not necessarily a bad thing!* In fact, it is a positive factor in this circuit since it makes the circuit flexible with a faster response speed.

Referring back to Figure 3.2, it can be seen that the jittery clock is caused by the multiple types of cycles used. This characteristic is confirmed by laboratory measurement. Figure 3.12 shows a period distribution measurement of a time-average-frequency clock signal. In this case, $\Delta = 62.5$ ps, $T_A = 8\Delta$, and $T_B = 9\Delta$. The weighting factor $r = 0.125$. As marked in the measurement, there are two peaks in the distribution corresponding to T_A and T_B, respectively.

When Time-Average-Frequency clock signal is used, the benefit of this "jittery" clock is its capability of generating more frequencies and enabling faster frequency switching. This jittery clock will be the key for system-level innovation, as will be discussed in Chapter 5.

3.7   FREQUENCY SWITCHING AND WAVEFORM ESTABLISHMENT IN TAF-DPS CLOCK GENERATOR

Since the TAF-DPS circuit constructs its waveform directly, it can switch its output frequency at a very high speed compared to the case of a PLL. This point is illustrated in Figure 3.13. It is demonstrated in the literature [Xiu12a] that it only takes two clock cycles, after the switching command is received, for a TAF-DPS clock to update its waveform. It is therefore claimed that the frequency switching speed is two clock cycles. This is easy to understand for the case of frequency switching that only involves a change in the integer part (refer to Figures 3.8 and 3.9 for evidence). However, more explanation is needed when the fractional part is also involved in the process. In such a case, it is necessary to distinguish between "a new frequency becomes effective" and "a new waveform is completed." In a conventional frequency wherein all the cycles have the same length-in-time, the two are identical since a particular value of length-in-time corresponds to a unique frequency. This is not the case for a time-average-frequency-based clock signal because, even with fixed T_A and T_B, the $f_{TAF} = 1/T_{TAF}$ can be different for different weighting factors. Therefore, it requires a time window of T_{FD} [the fundamental period; refer to (3.2)] to identify the value of f_{TAF} without ambiguity. Only after T_{FD} (at moment t_3 in Figure 3.13), from the moment of receiving the switch command, can the new waveform be completed. This scenario is illustrated on the right in Figure 3.13. It is important to point out that, although T_{FD} is the time required for a new waveform, the new frequency actually becomes effective after two clock cycles (the moment t_2). In other words, f_{TAF} will take a new value of f_2 if the TAF-DPS clock signal is measured immediately after t_2.

3.8   ALLAN VARIANCE OF TAF-DPS CLOCK SIGNAL

Allan variance $\sigma^2_y(\tau)$ is a measure of the frequency stability of clock signals and oscillators [All66, How99]. It is intended to estimate frequency stability due to noise processes. It does not measure frequency drift caused by systematic error or imperfections in system or induced by temperature variation. Allan variance converges for all noise processes observed in precision oscillators. It has a straightforward relationship to spectral density. It is faster and more accurate than the fast fourier transform

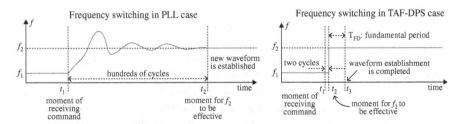

FIGURE 3.13   Frequency switching in PLL case (left) and TAF-DPS case (right).

in estimating noise processes. Allan deviation $\sigma_y(\tau)$ is the square root of the Allan variance. A TAF-DPS clock signal, although using two or more types of clock cycles (this fact makes it look like a systematic error or imperfection), does not introduce any frequency drift. As a matter of fact, the very purpose of using multiple types of cycles is to make the TAF-DPS clock's frequency converge to the designed value of f_{TAF}. Thus, Allan variance and Allan deviation are good tools to study the TAF-DPS clock signal further.

Figure 3.14 illustrates the parameters used to calculate the Allan variance. A oscillator, or a clock signal, under study can be described by (3.6), where $\Phi(t)$ is the total phase, f_n is the normal frequency, $f(t)$ is the frequency, and $\varphi(t)$ is the phase disturbance. Allan variance $\sigma^2_y(\tau)$ is obtained when the oscillator (clock signal) is compared against a reference oscillator of frequency f_n. The time error $x(t)$, when measured against this reference, can be expressed as in (3.7). Using f_n, the fractional frequency $y(t)$ can be defined and expressed as in (3.7) as well. When the Allan variance is calculated, an observation time τ is defined as illustrated in Figure 3.14. A time error series $x(i)$ is created by taking a sample at each τ. The average fractional frequency $\bar{y}(t, \tau)$ is defined over the observation time τ at time t, as shown in (3.8). The corresponding series $\bar{y}(i)$ can be generated subsequently. From this series of $\bar{y}(i)$, the Allan variance $\sigma^2_y(\tau)$ is calculated using (3.9), where M is the number of frequency samples. From the relationship of $\bar{y}(i)$ and $x(i)$ in (3.8), the Allan variance $\sigma^2_y(\tau)$ can also be calculated using time series $x(i)$ as shown in (3.9):

$$V(t) = A \, \sin[\Phi(t)] = A \, \sin[2\pi f_n t + \varphi(t)] \qquad \Phi(t) = 2\pi f_n t + \varphi(t) \qquad (3.6)$$

$$f(t) = \frac{1}{2\pi}\frac{d\Phi(t)}{dt}$$

$$x(t) = \frac{\varphi(t)}{2\pi f_n} = \frac{\Phi(t)}{2\pi f_n} - t \qquad y(t) = \frac{f(t) - f_n}{f_n} = \frac{f(t)}{f_n} - 1 \qquad (3.7)$$

$$\bar{y}(t, \tau) = \frac{1}{\tau}\int_0^\tau y(t + t_a)\, dt_a = \frac{x(t + \tau) - x(t)}{\tau} \qquad \bar{y}_i(t, \tau) = \frac{x_{i+1} - x_i}{\tau} \qquad (3.8)$$

$$\sigma^2(\tau, M) = \frac{1}{2(M-1)}\sum_{t=0}^{M-2}(\bar{y}_{i+1} - \bar{y}_i)^2 = \frac{1}{2\tau^2(M-1)}\sum_{i=0}^{M-2}(x_{i+2} - 2x_{i+1} + x_i)^2 \quad (3.9)$$

Based on (3.8) and (3.9), a tool is developed to calculate the Allan deviation $\sigma_y(\tau)$ for TAF-DPS clock signals. The first task accomplished by this tool is the confirmation that $\sigma_y(\tau) = 0$ for all the cases of frequency control words having no fractional part. This is because, when $r = 0$, the TAF-DPS clock signal is the conventional one where all the cycles have same length-in-time. When $r \neq 0$, the irregularity in the clock cycles' lengths will be regarded as "a type of noise" by this tool and the Allan variance will be calculated. It is however expected that $\sigma_y(\tau) = 0$ if the observation time τ takes a value of an integer multiple of the fundamental period T_{FD}. The reason is that the accumulated effects caused by the cycle irregularity are canceled out at the end of each T_{FD}. This predication is also confirmed by this tool.

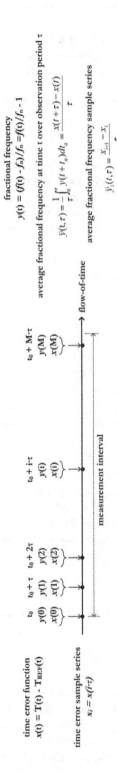

time error function
$x(t) = T(t) - T_{REF}(t)$

time error sample series
$x_i = x(i \cdot \tau)$

t_0 $t_0 + \tau$ $t_0 + 2\tau$ $t_0 + i \cdot \tau$ $t_0 + M \cdot \tau$
$y(0)$ $y(1)$ $y(2)$ $y(i)$ $y(M)$
$x(0)$ $x(1)$ $x(2)$ $x(i)$ $x(M)$

measurement interval

flow-of-time

fractional frequency
$y(t) = (f(t) - f_n)/f_n = f(t)/f_n - 1$

average fractional frequency at time t over observation period τ
$$\bar{y}(t,\tau) = \frac{1}{\tau} \int_0^\tau y(t + t_a) dt_a = \frac{x(t + \tau) - x(t)}{\tau}$$

average fractional frequency sample series
$$\bar{y}_i(t,\tau) = \frac{x_{i+1} - x_i}{\tau}$$

FIGURE 3.14 Time error sample and fractional frequency sample for calculating Allan variance.

For all the other cases, it is expected that the characteristics of this "noise" will be dependent on the frequency control word's integer I, the fraction r, and the base time unit Δ. Therefore, two approaches listed below are chosen to study their impacts:

1. Fix the base time unit Δ and vary the I and r (and thus f_{TAF}).
2. Fix the output frequency f_{TAF} and vary the base time unit Δ. This will force I and r to be changed indirectly.

In the generation of Allan plots, the observation times are chosen as 10 μs, 100 μs, 1 ms, 10 ms, 100 ms, and 1 s. Table 3.4 is the result of fixing the base time unit Δ and varying I and r for producing several frequencies. Figure 3.15 shows the corresponding Allan deviation plots. Table 3.5 is for the case of using several Δ values to produce a frequency of 1233.58 MHz. Figure 3.16 shows its corresponding Allan deviation plots.

From the study, several observations can be made as follows. These observations are made on data collected from many runs of simulation (Tables 3.4 and 3.5 are just examples):

- Allan deviation $\sigma_y(\tau) = 0$ for all the case of $r = 0$ (since there is no irregularity introduced by the T_A and T_B interleaved occurrence).
- Allan deviation $\sigma_y(\tau) = 0$ if observation time $\tau = nT_{FD} = nN_{TAF}T_{TAF}$, where n is an integer (because the irregularity introduced by the T_A and T_B interleaved occurrence is canceled out at the τ boundary).
- Allan deviation $\sigma_y(\tau)$ follows the trend of $1/\tau$ for all other cases (i.e., the effect of T_A and T_B interleaved occurrence is "white noise" when Allan deviation is concerned).
- Allan deviation $\sigma_y(\tau)$ does not have direct dependence on the size of I and r (since τ is usually several orders of magnitude larger than $I\Delta$).
- Allan deviation $\sigma_y(\tau)$ increases with Δ (since the time error increases with Δ).

One factor that makes time-average-frequency different from the often loosely used term "average frequency" is the existence of the fundamental period T_{FD}. Allan deviation $\sigma_y(nT_{FD}) = 0$ is the necessary condition for a clock signal to be qualified as a time-average-frequency clock.

3.9 BEHAVIOR OF TAF CLOCK SIGNAL UNDER THE SCOPE OF JITTER

The TAF-DPS mechanism of using two (or more) types of cycles to construct a clock signal is an intentional action. This is fundamentally different than the undesired phenomenon called jitter. It is, however, beneficial to study the TAF signal under the jitter scope since the tool of jitter has been widely used by engineers and researchers. Studying the TAF signal using a jitter scope can reveal some of its working details with connection to the established past knowledge. The three most often used jitter qualifiers are P2P period jitter, cycle-to-cycle (C2C) period jitter, and time interval

TABLE 3.4 Allan Deviation of Various Frequencies Using Base Time Unit $\Delta = 50$ ps

τ (s)	Freq. = 16.5, $I = 16, r = 0.5$, 1212.12 MHz	Freq. = 16.25, $I = 16, r = 0.25$, 1230.77 MHz	Freq. = 16.125, $I = 16, r = 0.125$, 1240.31 MHz	Freq. = 57.747, $I = 57, r = 0.747$, 346.34 MHz	Freq. = 16.213, $I = 16, r = 0.213$, 1233.58 MHz	Freq. = 162.13, $I = 162, r = 0.13$, 123.36 MHz
10^{-5}	3.00×10^{-6}	1.43×10^{-6}	3.25×10^{-6}	2.46×10^{-6}	3.47×10^{-6}	3.03×10^{-6}
0.0001	8.84×10^{-8}	2.50×10^{-7}	1.77×10^{-7}	3.09×10^{-7}	2.53×10^{-7}	3.27×10^{-7}
0.001	3.00×10^{-8}	3.00×10^{-8}	2.50×10^{-8}	3.16×10^{-8}	2.09×10^{-8}	3.49×10^{-8}
0.01	8.84×10^{-10}	1.13×10^{-9}	3.14×10^{-9}	7.75×10^{-10}	1.90×10^{-9}	1.77×10^{-9}
0.1	3.00×10^{-10}	2.50×10^{-10}	1.56×10^{-10}	1.36×10^{-10}	2.44×10^{-10}	3.31×10^{-10}
1	8.84×10^{-12}	2.42×10^{-11}	2.54×10^{-11}	3.54×10^{-11}	3.63×10^{-11}	3.98×10^{-12}

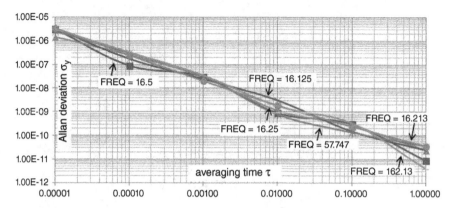

FIGURE 3.15   Allan deviation of various frequencies under same base time unit $\Delta = 50$ ps.

TABLE 3.5   Allan Deviation of Using Various Δ Values to Generate Frequency of 1233.58 MHz (0.81065 ns)

τ (s)	$\Delta = 25$ ps, Freq. = 32.426	$\Delta = 50$ ps, Freq. = 16.213	$\Delta = 75$ ps, Freq. = 10.8086667	$\Delta = 100$ ps, Freq. = 8.1065	$\Delta = 150$ ps, Freq. = 5.404333
1.00×10^{-5}	1.33×10^{-6}	3.47×10^{-6}	3.76×10^{-6}	5.51×10^{-6}	8.85×10^{-6}
0.0001	1.51×10^{-7}	2.53×10^{-7}	5.46×10^{-7}	6.35×10^{-7}	7.22×10^{-7}
0.001	1.25×10^{-8}	2.09×10^{-8}	4.96×10^{-8}	7.03×10^{-8}	8.89×10^{-8}
0.01	1.25×10^{-9}	1.90×10^{-9}	5.07×10^{-9}	6.61×10^{-9}	9.45×10^{-9}
0.1	1.08×10^{-10}	2.44×10^{-10}	3.75×10^{-10}	5.82×10^{-10}	7.50×10^{-10}
1	1.05×10^{-11}	3.63×10^{-11}	5.57×10^{-11}	5.27×10^{-11}	7.86×10^{-11}

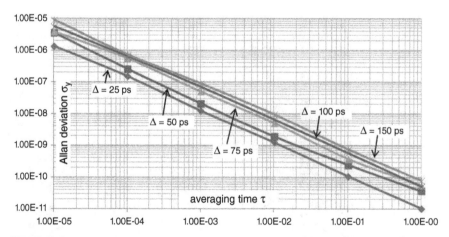

FIGURE 3.16   Allan deviation of using various Δ to generate frequency 1233.58 MHz (0.81065 ns).

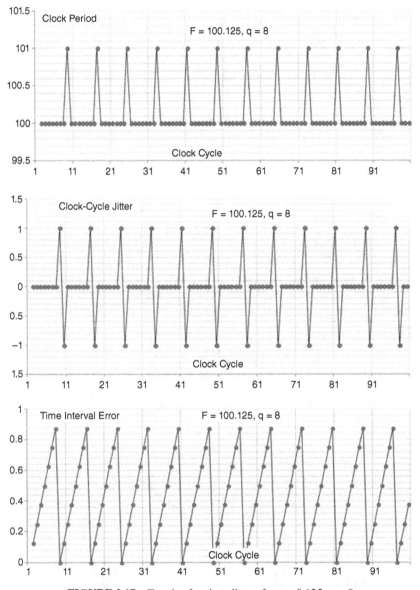

FIGURE 3.17   Trends of various jitters for $r = 0.125$, $q = 8$.

error (TIE). Figure 3.17 shows the three jitters for $F = 100.125$ [$N_{TAF} = q = 8$; refer to (3.1)]. All the Y-axises are in unit of Δ. For a TAF signal, its "jitter" is unambiguously determined by the fraction used in the control word. The top plot shows the trend of the clock period. From the plot, it is seen that the P2P period jitter is 1Δ. It is bounded regardless of the number of samples observed. This is because it is deterministic. From the middle plot, the C2C period jitter is also 1Δ and is bounded at this value. (In this case, the C2C jitter plot is calculated using a plurality

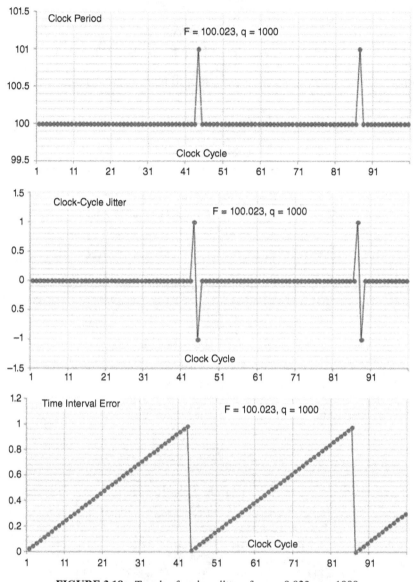

FIGURE 3.18   Trends of various jitters for $r = 0.023$, $q = 1000$.

of consecutive cycles. In C2C jitter, two consecutive cycles are required to make one C2C calculation.)

Being used for describing long-term jitter behavior, the TIE grows monotonically before each fractional overflow happens. This pattern could be quite complex depending on the value of the fraction. Further, it can repeat several times within each fundamental period T_{FD} [refer to (3.2)]. The TIE resets after a time period T_{FD} is passed, as illustrated in the plot at the bottom of Figure 3.17. Figure 3.18 shows the case $F = 100.023$ ($N_{TAF} = q = 1000$). In this case, the TIE will return to zero at

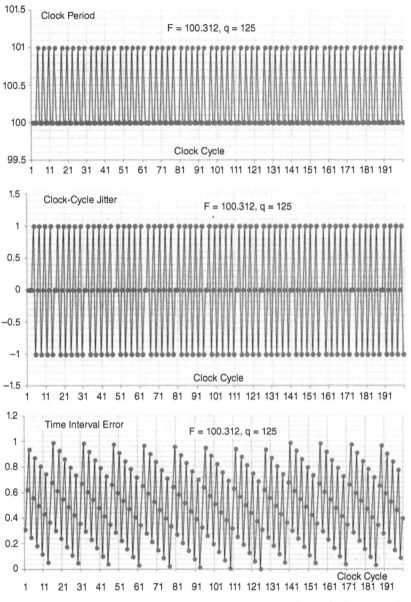

FIGURE 3.19 Trends of various jitters for $r = 0.312$, $q = 125$.

the 1000th cycle (not shown in the plot). Since $23/1000 \approx \frac{1}{40}$, there is a "semireset point" (local minimum) for about every 40 cycles. It is caused by the overflow from the fractional part. At these points, however, the TIE does not completely return to zero. Figure 3.19 is the case of $F = 100.312$ ($N_{TAF} = q = 125$). From this figure, it is seen that the TIE returns to zero every 125 cycles. There are several local minimums along the trend whose values approach zero gradually. Those TIE plots reveal the

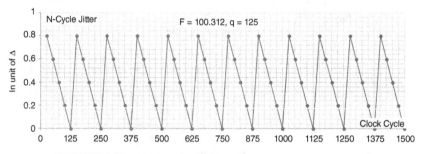

FIGURE 3.20   Long-term N-cycle jitters for $r = 0.312$, $q = 125$.

fact that the "jitter" in the TAF signal does not accumulate. Figure 3.20 shows the long-term N-cycle jitter for the case $F = 100.312$. Again, it shows that for a TAF signal the jitter does not accumulate. Further, if $N = nN_{TAF}$, where n is an integer, the N-cycle jitter is zero.

Three important points can be summarized:

- All "jitters" in the TAF signal are bounded. They do not accumulate.
- The "jitters" in the TAF signal are well behaved. They are predictable.
- For a TAF signal, if necessary, the jitter's behavior can be altered to produce beneficial results.

3.10   SPECTRUM OF TAF CLOCK SIGNAL

Table 3.6 lists the parameters related to the spectrum analysis of a TAF clock signal. Due to the use of multiple types of pulses in the clock pulse train, the spectrum of a TAF clock signal is fairly complex. Compared to the ideal 50% duty-cycle square-wave clock signal (the conventional frequency clock signal) where the fundamental frequency is the clock frequency f, the TAF clock signal fundamental frequency is not the time-average-frequency f_{TAF}. It is rigorously proven, however, that the main tone of the TAF clock signal's spectrum lies in the location of f_{TAF} [Sot10a, Sot10b, Sot10c]. In other words, most of the clock energy is concentrated at the desired frequency (the designed target of the time-average-frequency clock pulse train). A rigorous mathematical analysis is presented in Section 5.4 of [Xiu12a]. The interested reader is referred to that work for further study.

It is pointing out that, because of the use of multiple types of pulses, the spectrum of the TAF clock signal can be manipulated with great flexibility. Its spectrum is very programmable. This could be useful for certain applications.

3.11   IMPACT OF IMPLEMENTATION IMPERFECTION

As illustrated in Figure 3.3, the mechanism of direct period synthesis is to count the number of base time units Δ as time flows forward and consequently generate

TABLE 3.6   Parameters for Spectrum Analysis on Time-Average-Frequency Clock Signal

Parameter	Definition	Comment
Δ	Base time unit	Absolute time measured in seconds
T_j	$T_j = I_j\Delta$, I_j is an integer.	Length-in-time of each individual pulse
Segment	Minimum section of waveform that is repeatable	Pulse train made of a plurality of pulses
N_{TAF}	Number of pulses in the segment	
T_{FD}	$T_{FD} = \Sigma\, T_j, j = 1, ..., N_{TAF}$	Fundamental period
T_{TAF}	$T_{TAF} = \Sigma\, a_i \cdot T_i, i = 1, ... n$ $(= T_{FD}/N_{TAF})$	Time-average period, n is the number of types of pulses in the TAF signal, a_i is the possibility of occurrence of type T_i
f_{FD}	$f_{FD} = 1/T_{FD}$	Fundamental frequency, shown as the space between the spurious tones in the spectrum.
f_{TAF}	$f_{TAF} = 1/T_{TAF} = N_{TAF} \cdot f_{FD}$	Time-average-frequency, the (N_{TAF})th harmonic of fundamental frequency, shown as the main tone in the spectrum.

the rising and falling edges at the predetermined counter outputs. Therefore, the key implementation imperfection having dominant impact on the clock output is the error on the base time unit Δ. As shown in Figure 3.4, Δ is realized as the time span between any two adjacent phases outputted from a PLL or a DLL or a divider chain. The mismatch on those phases leads to error on Δ. Before reaching the actual output pulse train generator, the toggle flip-flop, those phases could also be influenced by the mismatches of the delivery paths, the $K \rightarrow 1$ and $2 \rightarrow 1$ multiplexes. All those mismatches (the PLL/DLL/divider output, the multiplexes, the layout of the delivery paths, etc.) can be lumped together and modeled as the error on the base time unit Δ. Its impact on the output clock signal has been studied elsewhere [Xiu12b]. The key conclusions are listed below:

- The impact of the error on Δ (all the mismatches included) is deterministic. Once the mismatch has occurred, its characteristics are fixed.
- The error on Δ results in deterministic jitter on the output clock. This jitter does not accumulate. It is bounded.
- On the spectrum of the output clock, the error on Δ shows its impact as spurious tones (spurs). The location and magnitude of these spurs are fixed and calculable.

REFERENCES

[All66]   D. Allan, "Statistics of atomic frequency standards," *Proc. IEEE*, vol. 54, no. 2, pp. 221–230, 1966.

[All97]   D. W. Allan, N. Ashby, and C.C., Hodge, "The science of timekeeping," Application Note 1289, Hewlett-Packard, 1997. http://www.allanstime.com/Publications/DWA/Science_Timekeeping/TheScienceOfTimekeeping.pdf

[Bar02]   J. D. Barrow, *The Constants of Nature: From Alpha to Omega—The Numbers That Encode the Deepest Secrets of the Universe*, Pantheon Books, New York, 2002.

[Gui10]   P. Gui, C. W. Huang, and L. Xiu, "The effect of flying-adder clock on digital-to-analog converters," *IEEE Trans. Circuit Sys. II*, vol. 57, no. 1, pp. 1–5, Jan. 2010.

[How99]   D. A. Howe, "Total variance explained [in frequency stability]," paper presented at the Joint Meeting EFTF-IEEE, *Proc. 1999 IEEE Int. Freq. Control Symp.*, vol. 2, pp. 1093–1099, 1999.

[Sot10a]   P. Sotiriadis, "Theory of flying-adder frequency synthesizers part I: Modeling, signals periods and output average frequency," *IEEE Trans. Circuits Syst. I*, vol. 57, no. 8, pp. 1935–1948, Aug. 2010.

[Sot10b]   P. Sotiriadis, "Theory of flying-adder frequency synthesizers part II: Time and frequency domain properties of the output signal," *IEEE Trans. Circuits Syst. I*, vol. 57, no. 8, pp. 1949–1963, Aug. 2010.

[Sot10c]   P. Sotiriadis, "Exact spectrum and time-domain output of flying-adder frequency synthesizers," *IEEE Trans. Ultrason., Ferromagnet. Freq. Control*, vol. 57, no. 9, pp. 1926–1935, Sept. 2010.

[Tal12]   S. Talwalkar, "Quantization error spectra structure of a DTC synthesizer via the DFT axis scaling property," *IEEE Trans. Circuits Syst. I*, vol. 59, no. 6, pp. 1242–1250, June 2012.

[Tal13]   S. Talwalkar, "Digital-to-time synthesizers: Separating delay line error spurs and quantization error spurs," *IEEE Trans. Circuits Syst. I*, vol. 60, no. 10, pp. 2597–2605, Oct. 2013.

[Xiu08a]   L. Xiu, "The concept of time-averaged frequency and mathematical analysis of flying-adder frequency synthesis architecture," *IEEE Circuit Syst. Mag.*, 3rd quarter, pp. 27–51, Sept. 2008.

[Xiu08b]   L. Xiu, "Some open issues associated with a new type of component: Digital-to-frequency converter," *IEEE Circuit Syst. Mag.*, 3rd quarter, pp. 90–94, Sept. 2008.

[Xiu10]   L. Xiu, C. W. Huang, and P. Gui, "The analysis of harmonic energy distribution portfolio for digital-to-frequency converters," *IEEE Trans. Instrum. Meas.*, vol. 59, pp. 2770–2778, Oct. 2010.

[Xiu11]   L. Xiu, M. Ling, and H. Jiang, "A storage based carry randomization techniques for spurs reduction in flying-adder digital-to-frequency converter," *IEEE Trans. Circuit Sys. II*, vol. 58, no. 6, pp. 326–330, June 2011.

[Xiu12a]   L. Xiu, *Nanometer Frequency Synthesis beyond Phase Locked Loop*, Wiley-IEEE Press, Hoboken, NJ, Aug. 2012.

[Xiu12b]   L. Xiu, K.-H. Lin, and M. Ling, "The impact of input-mismatch on flying-adder direct period synthesizer," *IEEE Trans. Circuit Syst. I*, vol. 59, pp. 1942–1951, Sept. 2012.

4

TIME-AVERAGE-FREQUENCY AND SPECIAL CLOCKING TECHNIQUES: GAPPED CLOCK, STRETCHABLE CLOCK, AND PAUSIBLE CLOCK

4.1 GAPPED CLOCK AND SYNCHRONOUS FIFO

Gapped clocking is a commonly used technique in optical transport networks (OTNs), broadcast video, and other applications [Alt13, Max02, Mic10, Sil10]. Its purpose is to compensate the transmission rate difference between the line clock and the client clock (in the case of OTNs). In such applications, for the purpose of framing, justification, forward error correction (FEC), etc., additional bits are added to the raw data to facilitate the protocol implementation in the process of data transmission. When the data packet is received, these extra bits are stripped. The gapped clock is used to make up the difference between the transmission rate and the payload rate by periodically removing pulses in the higher frequency clock. Figure 4.1 shows the mechanism of gapped clock. In the top trace, one pulse is removed from the shown portion of the clock pulse train. In practice, multiple pulses can be deleted in one shot if needed. When the gapped clock is used in transmitting data, the data associated with the missing pulse are prolonged (extended one or more cycles). When data transmitted by a gapped clock are received by a receiver driven by a normal clock, the data associated with missing pulses are invalid. In summary, when a gapped clock is involved in data transmission, the number of bits in the data has to equal the number of pulses in the clock pulse train. No new data should be created for the missing pulse.

Figure 4.2 depicts the architecture of a gapped clock in a transmitter [Sil10]. Raw data are generated by a clock of frequency f_l. Before it can be serialized and sent to the transmission channel, additional bits have to be added by a frame processor

From Frequency to Time-Average-Frequency: A Paradigm Shift in the Design of Electronic Systems, First Edition. Liming Xiu.
© 2015 The Institute of Electrical and Electronics Engineers, Inc. Published 2015 by John Wiley & Sons, Inc.

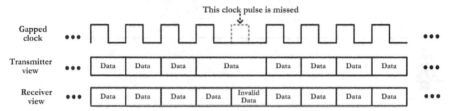

FIGURE 4.1   Gapped clock and its impact on transmitter and receiver.

(mapper) which is driven by a clock of frequency f_O. The frame processor adds protocol-supporting information into the original data stream. Since more bits are outputted from the mapper, we have $f_O > f_I$. Thus, a FIFO is placed in between the data source and the mapper to handle the rate difference. The FIFO is synchronous with clock ports on both sides. If no compensation is performed (i.e., using CLK_O of f_O to drive the FIFO's CK_{out} port directly), the FIFO will eventually underflow. To prevent this scenario from happening, based on the FIFO status (such as underflow, overflow, half full, almost full, and almost empty), the gapping logic deletes pulses from the CLK_O pulse train systematically. As a result, the number of transition edges embedded in the gapped clock CLK_G is averagely equal to that of CLK_I so that the inflow and the outflow of the FIFO are matched over a specified time span. This goal can be expressed as in (4.1), where N_{CLKI} and N_{CLKG} are the numbers of transition edges (assuming all circuits are single-edge triggered) in clocks CLK_I and CLK_G, respectively. The equation is established in a specified time span. The time span is inversely proportional to the required size of the FIFO:

$$N_{\text{CLKG}} = N_{\text{CLKI}} \quad \text{in a specified time span} \tag{4.1}$$

Figure 4.3 shows the architecture of a gapped clock in a receiver [Sil10]. Serial data received from the transmission channel are first processed by a clock data recovery (CDR) circuit and the incoming data stream is separated into recovered data and recovered clock. The recovered data are then processed by the demapper so that the auxiliary bits can be removed. As a result, there are less bits in the output side of the demapper than the input side. This goal is also achieved by the gapped clock approach of removing pulses so that fewer transitions result. Unlike the previous case where the gapped clock is only used to drive the FIFO, this rate-modified clock also needs to drive a downstream processor, such as a deserializer. For this reason, an additional requirement in this case is to smooth the gapped clock before sending it to the downstream processor. This is because the missed pulses in the gapped clock are considered to cause "jitter" in clock CLK_G. To deal with this issue, a jitter clean PLL is usually used to attenuate the effect caused by the missed pulses. The purpose of this PLL is to block the low-frequency modulation caused by the missing pulses and make (4.2) valid, where N_{CLKO} and N_{CLKG} are the numbers of transition edges in clocks CLK_O and CLK_G, respectively:

$$N_{\text{CLKO}} = N_{\text{CLKG}} \quad \text{in a specified time span} \tag{4.2}$$

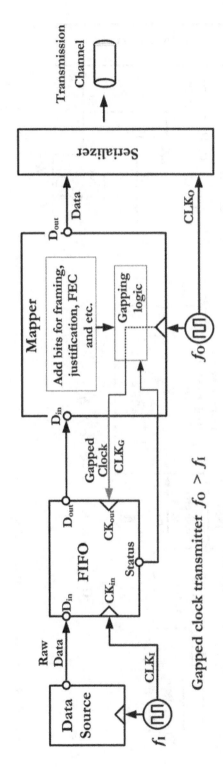

FIGURE 4.2 Architecture of gapped clock transmitter.

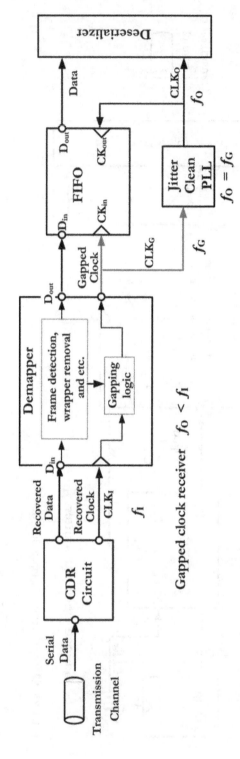

FIGURE 4.3 Architecture of gapped clock receiver.

4.2   STRETCHABLE CLOCK, PAUSIBLE CLOCK, AND ASYNCHRONOUS FIFO

An important trend in today's SoC design is heterogeneous clocking. This is due to the fact that today's large systems are heterogeneous and have many functional blocks running at different frequencies. The major challenge in a heterogeneous system is the data communication between these blocks. Clocking is the key issue in the design of a communication scheme in such an environment [Yun96, Bor97, Mul05]. When communication between blocks of different clock frequencies is required, FIFO is used between the blocks to facilitate the rate adjustment. The architectures presented in Figures 4.2 and 4.3 use a combination of a gapped clock and FIFO to achieve the communication. The two systems at the FIFO's left and right sides are synchronous. The FIFO in this case is also synchronous. Usually, the data rates of these systems at the two sides of FIFO are different but both are fixed and known. The gapped clock is produced by removing pulses from a clock signal generated by an "untouchable" clock source (in other words, the gapped clock is created outside the clock generator box). From a circuit design point of view, this is a relatively easy task since it does not require analog design skill, which most system designers do not possess.

There are systems, however, in which the clock rate at either (or both) side(s) of the FIFO is variable and/or unknown. In this scenario, handshake signals are usually provided to assist the communication, and the interfaces between the FIFO and the functional blocks are asynchronous. In other words, although the functional blocks are synchronous, the FIFO is asynchronous. The essence of using handshake signals is to ensure that data must be ready and reliable before the arrival of the receiving clock edge. This is the so-called value-safe method that avoids the metastability problem [Cha84]. When this requirement is implemented in circuit design, it is necessary that the clock signal only rise (assuming a rising edge clock) after data become stable. Or, the clock edge must be delayed (paused or stretched) if necessary. For a clock, this task requires the modification of the clock generator circuit. The example depicted in Figure 4.4 demonstrates this.

In the architecture depicted on the left in Figure 4.4, a FIFO is used to facilitate the data communication between the TX and RX, which are driven by clock signals CLKT and CLKR, respectively. The FIFO is an asynchronous design which does not have a clock port. Instead, it has ports {PUT, OK_to_PUT}, {TAKE, OK_to_TAKE} on each side of the FIFO for the handshake. The signals {TX_DATA, READY_to_PUT} and {RX_DATA, READY_to_TAKE} are associated with the TX and RX blocks, respectively. Clock signals CLKT and CLKR are generated from two local generators. These clock generators are constructed from a complementary metal–oxide–semiconductor (CMOS) ring oscillator with an enable signal EN that can disable the oscillation. Thus, CLKT and CLKR can be paused or stretched by the handshake signals OK_to_PUT and OK_to_TAKE through EN ports. The waveforms displayed on the right side show the mechanism of handshaking and the way that the clock signals are modified.

At first, CLKT starts normally and is sent to the TX block. When TX_DATA is ready to be sent to RX (through the FIFO), the TX asserts the READY_to_PUT by

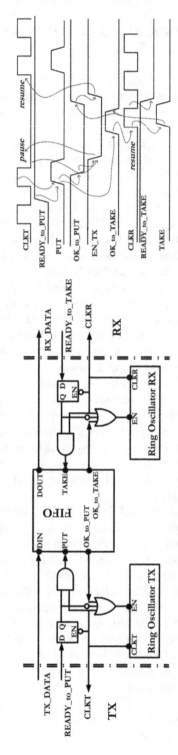

FIGURE 4.4 Pausible or stretchable clock and asynchronous interface (left) and communication handshaking (right).

raising its level. At the low portion of CLKT, READY_to_PUT is passed through the latch and, at the next high portion of CLKT, the port PUT receives a true signal (level high) which signals the FIFO that everything is ready from the TX side. After receiving this signal at PUT, FIFO immediately rejects OK_to_PUT by lowering its level. This will pause the CLKT by disabling the EN signal of the clock generator for TX. At the same time, the FIFO latches the data though its DIN port. Then, it asserts the OK_to_TAKE after the data are stored in FIFO reliably and are ready to be taken. This OK_to_TAKE signal enables the RX clock generator that resumes the CLKR signal, which is paused before and during the PUT operation. As a result, the data inside FIFO are taken out by RX though port DOUT and the port TAKE has an asserted signal which indicates that the data have been taken. The TAKE signal also raises the OK_to_PUT, which resumes the CLKT in the TX clock generator by enabling the oscillator. At this time, a full cycle of data in and data out is accomplished within FIFO. From this discussion, it is clear that the two signals OK_to_PUT and OK_to_TAKE control the two clock generators and modify the clock waveforms. During this process, the clock signals are paused or stretched to accommodate the date operation. The stretchable and pausible clock will be discussed again in later chapter in connection with the application environment of a NoC (network-on-chip).

Another usage of the stretchable clock and pausible clock is to facilitate the DVFS (dynamical voltage and frequency scaling), a chip implementation technique for low-power operation. It dynamically adjusts the power supply level and/or operating frequency based on work loading. During the time window that the power level is switched, the clock signal has to be paused or stretched.

4.3 TIME-AVERAGE-FREQUENCY OPERATION OF GAPPED CLOCK, STRETCHABLE CLOCK, AND PAUSIBLE CLOCK

The techniques discussed so far in this chapter, including the gapped clock, stretchable clock, pausible clock, and possible others all have the Time-Average-Frequency concept work behind the scene. All these special clocking techniques illustrate the fact that the essence of clock frequency is the accomplishment of a specified number of operations within the time window of 1 s. As long as the specified number of operations is successfully executed (ensured by the specified number of transitions/edges in a clock pulse train), the system is not concerned with the detail structure of the clock pulse train. In this regard, the clock signal of the gapped clock, stretchable clock, and pausible clock does not differ from that generated by TAF-DPS. The TAF-DPS is a smoother, or less severe, operation than these special clocking techniques since, instead of modifying a particular pulse train by removing one or more whole pulse(s), TAF-DPS modifies a clock pulse by only a small fraction of the pulse length. The following example demonstrates this point.

In [Sil10], a 100-MHz clock is used to illustrate the gapped clock. For every 100 pulses of this clock signal (time window of 1 μs), one pulse is removed from the train. This results in 99 million pulses in the time window of 1 s. It is measured as 99 MHz if a frequency counter is used. Using the time-average-frequency concept of

TABLE 4.1   Comparison of TAF-DPS and Gapped Clock in Generating 99 MHz

	Number of Pulses	T_B/T_A	Weighting r	Base Time Unit Δ	Clock Spectrum
TAF-DPS	99×10^6	$51/50 = 1.02$	0.5050505051	0.2 ns	Main tone at 99 MHz
Gapped clock	99×10^6	$2/1 = 2$	0.0101010101	10 ns	Main tone at 100 MHz

Chapter 3, this gapped clock can be regarded as using two types of cycles $T_A = 10$ ns, $T_B = 20$ ns ($\Delta = 10$ ns) and $r = 1/99 = 0.01010101010101$ ($N_{\text{TAF}} = 99$).

We now consider using TAF-DPS to generate this clock of 99 MHz. If a base unit $\Delta = 0.2$ ns is created from a reference clock of 156.25 MHz and using $K = 32$ ($\Delta = 1/(Kf_r)$), the TAF-DPS can generate the 99 MHz clock frequency through this configuration: $T_A = 50\Delta = 10$ ns, $T_B = 51\Delta = 10.2$ ns, and $r = 0.5050505051$ (=100/198). Table 4.1 lists the key characteristics of these two approaches. It is worth mentioning that, depending on the available reference frequency, there are many other TAF-DPS configurations that can be chosen to generate the 99 MHz.

Figure 4.5 shows the trend of pulse length vs. time. It is obtained from transistor-level simulation. The traces are the results of direct measurement of a simulated clock waveform. In the left plot, the trend of a normal 100-MHz clock is displayed. The simulation time is 10 μs, which covers about 1000 pulses. As shown, all the pulses have the same 10 ns length. In the middle left plot, the trend associated with the gapped clock is displayed. Since one pulse is removed from every 100 pulses, two types of pulse lengths result: 10 and 20 ns. This happens every 1 μs. The trend of the TAF-DPS clock is displayed in the middle right. Here there are two types of pulses as well: 10 and 10.2 ns. Compared to the gapped clock case, these two types of pulses are much closer to each other in size. But the T_B type occurs much more frequently. It happens about once every two cycles since r is roughly 0.5 (for every 198 pulses, exactly 100 pulses are extended from T_A to T_B, which is 1 $\Delta = 0.2$ ns longer). A zoom-in view is provided to reveal this fact. The plot on the right displays the three traces together but in a much smaller time span of 200 ns. The level of severity in the pulse length's modulation can be more visually appreciated.

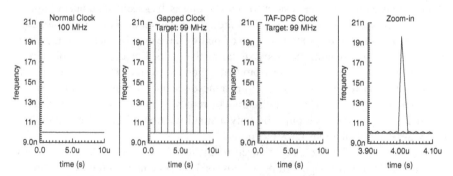

FIGURE 4.5   Trend of pulse length vs. time: normal clock of 100 MHz (left), gapped clock (middle left), TAF-DPS clock (middle right), and zoom-in view (right).

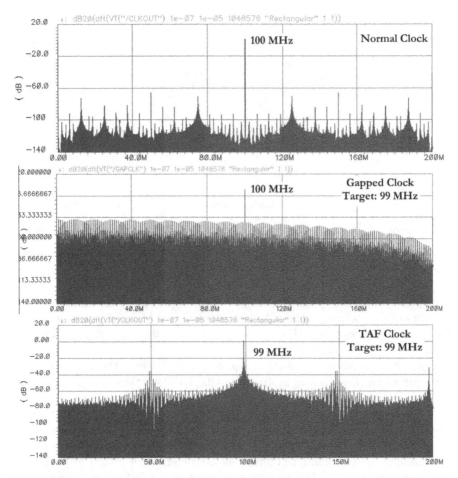

FIGURE 4.6 Clock spectra of 100-MHz clock (top), 99-MHz gapped clock (middle), and 99-MHz TAF-DPS clock (bottom).

Figure 4.6 shows the clock spectra of the 100 MHz clock, 99 MHz gapped clock, and 99 MHz TAF-DPS clock, also obtained from transistor-level simulation with a simulation time of 10 μs. The three spectra are plotted using the same scale in both the X- and Y-axes for easy comparison. In the 100 MHz normal clock (top trace), the main tone is exactly at 100 MHz, as expected. In the gapped clock (middle trace), the main tone is not at 99 MHz but still at 100 MHz with strong spurious tones spaced at 1 MHz. This is caused by the 1 MHz modulation of periodically deleting a pulse once every 100 pulses. The TAF-DPS output is displayed at the bottom. In this case the main tone is exactly at 99 MHz (the intended target). The spurious tones caused by fraction 0.50505051 are weaker in strength (compared to those in the gapped clock). Since this fraction of 100/198 cannot be represented in the binary system with 100% precision, the resultant spectrum bears some complexity. However, the dominant

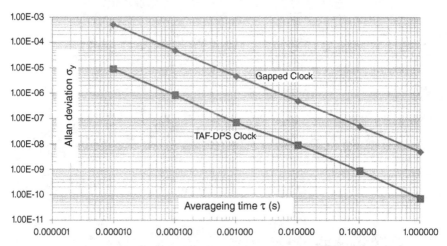

FIGURE 4.7 Allan deviation plot of gapped clock (top trace) and TAF-DPS clock (bottom trace).

modulation pattern of $(^{1}/_{2})$99 MHz is clearly visible since 100/198 (=0.5050505051) is close to $^{1}/_{2}$. It is worth mentioning that, besides being obtained from simulation, this spectrum can be calculated though analysis using the theory developed elsewhere [Sot10a, Sot10b, Sot10c, Xiu12].

Figure 4.7 presents the Allan variance plots of the two cases. The Allan deviation of the gapped clock is much larger than that of the TAP-DPS clock since its time base unit $\Delta = 10$ ns is significantly larger ($\Delta = 0.2$ ns for TAF-DPS).

When the gapped clock, stretchable clock, and pausible clock are viewed under the concept of time-average-frequency, it is interesting to point out that the possibility of occurrence of the different cycles (the fraction r in the case of two types of cycles) is determined dynamically in real time. Operationwise, this is achieved by the data flow adjustment inside the FIFO. Continuous update of the FIFO status and the clock generator's pulse length modification form a feedback loop. The value of r is then determined implicitly. In the TAF-DPS approach, the same effect can be achieved by adjusting r dynamically in an explicit way. Over a relatively larger time span, the average of r's many instantaneous values is the intended value that defines the actual data rate.

An important topic related to the gapped clock, pausible clock, and stretched clock is clock jitter. It is crucial to understand that the pulse-length modification introduced by these special clocking techniques is an intended operation. Its impact is known and deterministic. This is completely different than the inherent spirit embedded in the concept of jitter, whose essence is random, or deterministic but unintended. Although this issue of jitter is not discussed explicitly by [Cha84, Bor97, Yun96, Tee07], it is evident from the architectures depicted in Figures 4.2, 4.3, and 4.4 that the pulse-length modifications in these special clocking techniques are not treated as jitter by these researchers, with which this author agrees. In these systems, the gapped clocks are used to drive FIFO directly; the pausible and stretched clock is

TABLE 4.2   Special Clocking Techniques of Modifying Pulse Length

	Usage	FIFO Type	Severity of Pulse Modification	Time-Average-Frequency Concept	Jitter View
TAF-DPS	Frequency synthesis, data communication	Synchronous	Low	Used explicitly	Sometime mistakenly regarded as jitter
Gapped clock	Data communication of fixed and known rates	Synchronous	High	Used implicitly	Always mistakenly regarded as jitter
Pausible/ stretchable clock	Data communication of variable or unknown rates DVFS	Asynchronous	High	Used implicitly	Not viewed as jitter
Clock gating	Low-power operation	N/A	Very high	Used implicitly	Not viewed as jitter at all

used to drive the TX and RX directly. As demonstrated in Chapter 3, as long as the FIFO or other digital processing blocks is setup constrained by the smallest pulse length, it is absolutely safe to use such a clock to drive the circuit. For this reason, it is suggested that the jitter clean PLL in Figure 4.3 is not needed since the deserializer is a digital processing block. As a "gentler cousin" of these special clocking techniques, the TAF-DPS clock is surely suitable for such applications. Its application in data communication will be discussed in depth in a later chapter. Table 4.2 summarizes the discussion. In this table, clock gating is also listed as one of the clocking techniques of modifying pulse length. This is true since the clock-quiet period can be considered as a T_B type of cycle, whose size is very large compared to the normal cycle (the T_A type). In principle, the pulse-length modification introduced by clock gating, which has never been considered as jitter, is not different from that created by the TAF-DPS clock, gapped clock, and pausible and stretchable clocks.

REFERENCES

GAPPED CLOCK

[Alt13]   Altera, "Clock reconstruction with low-cost external DCXO," Application Note AN-695, Altera Corp., 2013. http://www.altera.com/literature/an/an_695.pdf

[Max02]   Maxim, "DS31256 gapped clock applications," Application Note 392, Maxim Integrated, 2002.

[Mic10]   Microsemi, "Optical transport network," Application Note, Microsemi Corp., 2010.

[Sil010]   Silicon Labs, "Introduction to gapped clocks and PLLs," Application Note AN561, Silicon Labs, 2010.

PAUSIBLE AND STRETCHABLE CLOCK

[Bor97]   D. S. Bormann and P. Y. K. Cheung, "Asynchronous wrapper for heterogeneous system," *ICCD'97 Proc.*, pp. 307–314, 1997.

[Cha84]   D. M. Chapiro, "Globally-asynchronous locally-synchronous systems," Ph.D. Thesis, Stanford University, Stanford, CA, Oct. 1984.

[Mul05]   R. Mullins, "Asynchronous vs. synchronous design techniques for NoCs," Tutorial, International Symposium on System-on-Chip, Tampere, Finland, Nov. 2005. http://www.cl.cam.ac.uk/~rdm34/soc2005tutorial.pdf

[Sot10a]   P. Sotiriadis, "Theory of flying-adder frequency synthesizers part I: Modeling, signals periods and output average frequency," *IEEE Trans. Circuits Syst. I*, vol. 57, no. 8, pp. 1935–1948, Aug. 2010.

[Sot10b]   P. Sotiriadis, "Theory of flying-adder frequency synthesizers part II: Time and frequency domain properties of the output signal," *IEEE Trans. Circuits Syst. I*, vol. 57, no. 8, pp. 1949–1963, Aug. 2010.

[Sot10c]   P. Sotiriadis, "Exact spectrum and time-domain output of flying-adder frequency synthesizers", *IEEE Trans. Ultrason. Ferromagnet. Freq. Control*, vol. 57, no. 9, pp. 1926–1935, Sept. 2010.

[Tee07]   P. Teehan et al., "A survey and taxonomy of GALS design styles," *IEEE Design & Test of Computers*, vol. 24, no. 5, pp. 418–428, Sept.-Oct. 2007.

[Xiu12]   L. Xiu, *Nanometer Frequency Synthesis beyond Phase Locked Loop*, Wiley-IEEE Press, Hoboken, NJ, Aug. 2012.

[Yun96]   K. Y. Yun and R. P. Donohue, "Pausible clocking: A first step toward heterogeneous system," *ICCD'96 Proc.*, pp. 118–123, 1996.

5

MICROELECTRONIC SYSTEM DESIGN IN THE FIELD OF TIME-AVERAGE-FREQUENCY: A PARADIGM SHIFT

The word *paradigm* is defined as a framework containing the basic assumptions, ways of thinking, and methodology that are commonly accepted by members of a scientific community. In his 1962 influential book *The Structure of Scientific Revolutions* [Kuh96], Thomas Kuhn describes the term *paradigm shift* as a change in the basic assumptions (the paradigms) within the ruling theory of science. Today, paradigm shift is used widely in both the scientific and nonscientific communities to describe a profound change in a fundamental model or perception of events. Since the clock concept was introduced into electronic system design many decades ago, it has been assumed by all engineers and researchers in the field that all the cycles in a clock pulse train have to be equal in length. This restriction on the clock signal (the paradigm) is helpful when it is used for driving electronic circuits. Since the fine structure of such a clock signal is simple, clock users have one less thing to worry about when they design electronic systems that use the clock signal as the driver. The reason that this form of clock signal has dominated electronic system design for such a long time is that, up to today, the requirement for IC clocking is mostly straightforward. A clock signal with a fixed rate is sufficient for most systems. However, the complexity of future systems changes the game. Low-power operation, low electromagnetic radiation, and complicated data communication schemes all require a clock signal that is flexible. Specifically, we want a clock signal whose frequency can (1) be arbitrarily set (within a small frequency granularity, similar to the fact that voltage level can be arbitrarily reached within a quantization resolution) and (2) be changed from one to another

From Frequency to Time-Average-Frequency: A Paradigm Shift in the Design of Electronic Systems, First Edition. Liming Xiu.
© 2015 The Institute of Electrical and Electronics Engineers, Inc. Published 2015 by John Wiley & Sons, Inc.

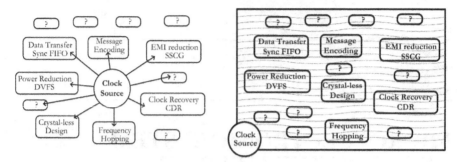

FIGURE 5.1 The view of using clock to solve various problems (left) and the view of building system in the field of "Time-Average-Frequency" (right).

quickly (similar to the fact that voltage level can make a transition quickly). Further, we want these two features to be available to a clock user at a reasonable cost.

This new demand for a flexible clock signal forces us to reinvestigate the concept of clock frequency (clock rate). From the discussion of Chapters 1 and 2, it is clear that the essence of clock frequency is to gauge the pace of events happening inside the electronic world. This task is executed by using each pulse's level-crossing-a-threshold to trigger events. Viewing from this perspective, the restriction that all cycles have the same length is not a fundamental necessity. It is "negotiable." This presupposition can be removed if it frees our hand in making better clockwork and ultimately making better systems. In Chapter 6 of [Xiu12a], "The New Frontier in Electronic System Design," several cases of using the Flying-Adder frequency synthesizer to enable system-level innovations are demonstrated. The approach used in that study is to solve various system-level problems by adopting a new type of clock generator. This point of view is illustrated on the left in Figure 5.1. In this methodology, the clock generator is part of the problem being investigated. In other words, the clock generator and the circuit it drives are an integrated system. However, this issue can be viewed from another angle. As discussed in Section 1.4, all electronic systems are frequency driven. Using this understanding, *an electronic system can be recognized as a machine operating within "a frequency field."* This field is created by an on-chip clock source. The purposes of introducing this concept of a frequency field are as follows:

- Make system-level issues stand out in the forefront and move the clock generation problem to the background.
- Establish the viewpoint that clock frequency is flexible (the field is dynamic; it constantly changes).
- Emphasize the fact that all the things in an electronic system are influenced by the clock (the field is felt by all the devices in it).

Following this line of thought, in this chapter, a new perspective is introduced: All system-levels problems are investigated within the time-average-frequency field.

This field is flexible: Its available frequencies are ample and the switching between frequencies is fast. This vision is depicted on the right-hand side of Figure 5.1. In this regard, using time-average-frequency to drive electronic system is a paradigm shift. It is a change of mindset.

5.1 DIGITAL DATA COMMUNICATION IN THE FIELD OF FREQUENCY

$$\text{CLK}(t) = p([(f + \delta f)t + \Phi(t)] \bmod 1), \qquad \frac{d\Phi(t)}{dt} = 0$$

$$p(t) = \begin{cases} 1 & 0 \le t < 0.5 \\ 0 & 0.5 \le t \le 1 \end{cases} \qquad\qquad (5.1)$$

$$f_{\text{avg}}(t) = f + \delta f + \frac{d\Phi(t)}{dt} = f + \delta f$$

Digital communication is controlled by a clock (more precisely, by its frequency). The left drawing in Figure 5.2. depicts the communication architecture in general. The transmitter (TX) and receiver (RX) are driven by their local clocks CLKT and CLKR, respectively. These clock signals are time variant. Their behaviors can be mathematically described using (5.1), where $p(t)$ is a pulse function, f is nominal frequency, δf is frequency offset, $\Phi(t)$ is the instantaneous phase whose first derivative has a mean value of zero, and f_{avg} is the average frequency [Mes90].

The key focus in digital data communication is data, which are driven by the clock and characterized by the data rate. The signal (digital data) can be classified based on the clock that drives it. In digital data communication, there are two types of digital signals: isochronous and anisochronous (middle drawing of Figure 5.2.). If $f + \delta f$ is a constant, the signal is said to be isochronous. For an isochronous signal, it is generally assumed that $\Phi(t)$ is bounded. In other words, an isochronous signal is a signal in which the time interval separating any two significant instants is equal to a unit interval or a multiple of the unit interval. Variation in the time interval is constrained within a specified limit. An anisochronous signal is one in which the time interval separating any two corresponding transitions is not necessarily related to the time interval separating any other two transitions.

When two or more signals are studied together, such as in the environment of communication depicted on the left in Figure 5.2, the relationship between the signals

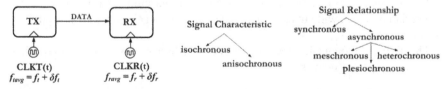

FIGURE 5.2   Digital data communication architecture (left); signal characteristic (middle); signal relationship (right).

in the TX and RX can be described as synchronous or asynchronous. Between the TX and RX, if their clock frequencies and offsets are the same ($f_t = f_r$, $\delta f_t = \delta f_r$), the two systems (more precisely the signals generated from the two systems) are said to be synchronous. A common example is the case where the data in both parties are controlled by the same clock. Any two systems that are not synchronous are asynchronous. For two signals with the same nominal frequency, the instantaneous phase difference between them can be expressed as

$$\delta\Phi(t) = (\delta f_1 - \delta f_2)t + [\Phi_1(t) - \Phi_2(t)] \tag{5.2}$$

Within the asynchronous approach, based on the δf characteristic, communication can be further categorized as mesochronous, plesiochronous, and heterochronous (refer to the right-hand side of Figure 5.2):

- Mesochronous   Two isochronous signals have exactly the same average frequency $f + \delta f$. Their phase difference $\delta\Phi$ is bounded. *Example:* Two signals are controlled by the same clock but a phase difference exists due to different delays.
- Plesiochronous   Two signals have average frequencies that are nominally the same but not exactly the same. Their phase difference can be expressed using (5.2). *Example:* Two signals are derived from independent oscillators that are supposed to have the same nominal frequency.
- Heterochronous   Two signals have different average frequencies.

As seen, both the signal characteristic and the signal relationship are defined through the driving clock. The quality of the clock source plays an important role in data communication. When the quality of a frequency source (clock source) is judged, three terms are often used: *accuracy*, *stability*, and *precision*. Accuracy is the extent to which a given measurement agrees with the definition of the quantity being measured. Stability describes the amount of change that occurs to certain thing as a function of some parameters such as time, temperature, and shock. Figure 5.3 illustrates four types of clock sources classified on its stability and accuracy defined around its oscillating frequency, where f_o is the target frequency. Precision is the extent to which a given measurement of one sample agrees with the mean of a measurement set that includes a multiple of measurements of the same sample. Figure 5.4 illustrates the difference of a frequency source's accuracy and precision with the help of a target used by a marksman. Finally, another term that is also of high importance is *reproducibility*. Reproducibility is the ability of a single frequency standard to produce, without adjustment, the same frequency each time it is put into operation [Vig01].

The TAF-DPS frequency (period) transfer function was expressed in equation (3.4) and is repeated here as

$$T_{\text{TAF}} = \frac{(qI + p)\Delta}{q} = F\Delta \quad \text{where } F = I + \frac{p}{q} = I + r \tag{5.3}$$

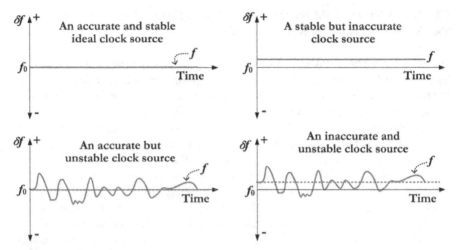

FIGURE 5.3 Quality of a clock source can be classified according to its stability and accuracy.

From this equation, it is clear that both the frequency accuracy and the precision of a clock source can be improved by the assistance of TAF-DPS. For example, the clock source can be made more accurate in its output frequency by adjusting the control word F (both I and r) toward its designed target. This is usually a one-time operation. On the other hand, the precision can be improved by fine turning r periodically to counteract the environmental change so that the output frequency is confined in a small range around the target [Xiu14a, Xiu14b].

The frequency field is generated by the clock source. The flexible TAF-DPS clock source enables a flexible time-average-frequency field. In the rest of this chapter, a variety of applications will be discussed around the concept of the time-average-frequency field. The intention is to help the reader establish a view of field, instead of just looking at the clock generator circuit. *In other words, it is desired for system designers to focus their attention on system-level issues, not the clock.* They can assume that the clock is flexible. This new view of investigation could lead to innovations because it frees the system designer's hand when various design options are investigated.

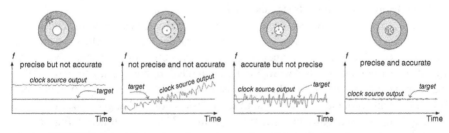

FIGURE 5.4 Illustration of accuracy and precision of a frequency source.

5.2 DATA TRANSFER IN TIME-AVERAGE-FREQUENCY FIELD: TRANSMITTING CLOCK AVAILABLE

Data transfer between blocks of different data flow rates is a very common task in VLSI system design. The transfer can be done asynchronously using handshake signals. But the most often used approach is the synchronous method where data on both sides are driven by clock signals of known frequencies. This approach is the mainstream method since there are many EDA (electronic design automatic) tools supporting this design methodology. In operation, a FIFO memory must be inserted between the blocks to facilitate the transfer. To match the rates of inflow and outflow, one or both sides of the clock signals have to be modified so that the average rates of the two sides are equal. The methods of clock modification include gapping the clock, pausing the clock, and stretching the clock. Strictly speaking, the systems driven by such types of clock signals are time-average-frequency systems, as discussed in Chapter 4.

As a general architecture, Figure 5.5. describes a communication scheme that uses TAF-DPS explicitly as an adaptive clock generator in the FIFO output side. As shown, the source data are generated in the CLK1 domain where the clock generator can be either a TAF-DPS source or a conventional source (such as a PLL). Its data rate is controlled by the clock frequency f_1. On the receiving side, the information is processed using the data rate controlled by CLK2 at f_2. There are many reasons why these two data rates could differ, such as redundant or auxiliary bits in the source data or additional information/bits needed in the CLK2 domain. The left drawing of Figure 5.5 depicts a scenario requiring different clock frequencies on the two sides of the FIFO. In the CLK1 domain, although the clock signal CLK1 is continuously running without interruption, the InData is not always presented in every cycle [Xiu07, Yan12]. The validity of the InData is indicated by an auxiliary signal, shown in the bottom trace. As a result, the actual data rate is lower than the clock frequency f_1. The instant data rate is shown in the trace labeled as "Data rate of InData." On the receiving side of the CLK2 domain, however, it is desired that valid data be associated with every clock cycle. This demands a clock frequency f_2 that is lower than f_1. Furthermore, to control the size of the FIFO from being too large, it is preferred that f_2 be adjusted dynamically according to the FIFO status. Thanks to its small frequency granularity and fast frequency switching, the TAF-DPS clock source (TAF synthesizer) can function as an adaptive clock generator, as illustrated in Figure 5.5. Based on the information of InData and the FIFO status, the adaptive clock generator can adjust its output clock frequency so that its average equals the InData rate, as shown in the trace labeled as "f_2 of CLK2."

Figure 5.6 provides a transistor-level simulation example that shows the characteristic of the adaptive clock generator. The bottom trace is the actual clock waveform. The top trace is its measured frequency. Based on the output of the calculation block, which is the frequency control word $F = I + r$ intended for configuring the TAF-DPS, the adaptive clock generator is able to adjust its output clock frequency in real time. It is important to point out that, when the control word is changed, the waveform transits from one to another in a glitch-free fashion. This is illustrated in Figure 5.7. As shown, the clock waveform changes seamlessly from one frequency to another.

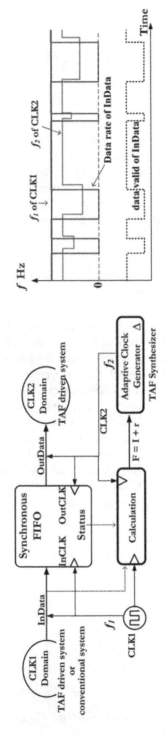

FIGURE 5.5 General data transfer architecture in TAF-driven system.

FIGURE 5.6   Adaptive clock generator's output: waveform (bottom); frequency (top).

Refer to Figure 3.8 for an example of seamless waveform transition obtained from lab measurement. The features demonstrated in Figures 3.5, 5.6, and 5.7 are the justification for the claim that the system is operating in a frequency field where the available rate is ample and the switching between rates is instantaneous. The TAF-DPS synthesizer is just the source to create this field. It is in the background. During the design process, it should not be the point of focus. Control of the data flow around the FIFO is the key focus.

5.3   DATA TRANSFER IN TIME-AVERAGE-FREQUENCY FIELD: CLOCKLESS TRANSMISSION

Section 5.2 discussed the data transfer case when the information regarding the transmitting clock is available to the receiver. In modern electronic system design, however, there is another important design approach wherein the information of the transmitting clock is not explicitly available to the receiver. Instead, the clock information is embedded in the data transmitted. This is the so-called clockless transmission that has become increasingly popular, especially for high-data-rate interchip communication. The advantage of this approach lies in the fact that it eliminates the skew problem between data and clock. As a result, on the receiver side, the embedded clock signal needs to be extracted by using the method of CDR. Figure 5.8 shows the general architecture of this clockless data transmission. Inside the receiver, clock CLKW is generated from the CDR circuit by extracting the clock from the data. It is then used to latch data. Before sending the received data to the downstream

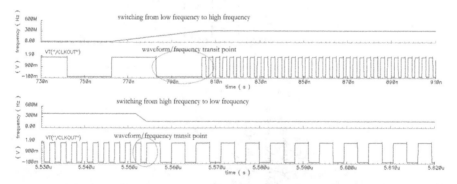

FIGURE 5.7   Glitch-free transition of clock waveform: fast to slow (bottom); slow to fast (top).

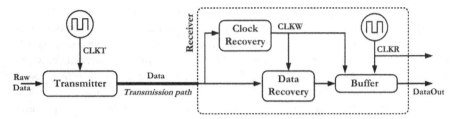

FIGURE 5.8   Architecture of clockless data communication.

processing unit, which is driven by clock CLKR, a buffer is inserted to accommodate the slight frequency difference between CLKW and CLKR. This difference could be caused by the removal of some auxiliary bits.

Figure 5.9 is a diagram of the buffer circuit. Storage cells are connected in a circular fashion. The write and read operations are controlled by write and read counters, respectively. The write and read pointers can only move one step at a time, in a designed direction between two adjacent cells. Based on the FIFO status, the CLKR frequency is adjusted from time to time so that the effect of bit removal can be compensated. The frequency is adjusted toward the goal of separating the two pointers by a distance of about half the buffer size. This ensures that the buffer does not overflow or underflow within a specified time frame. As can be understood from the discussion, the TAF-DPS is squarely fitted for this application due to its small frequency granularity and fast switching. When a time-average-frequency based synthesizer is used in CLKR, the system can be designed in such a way that the buffer would never overflow/underflow given a buffer of reasonable size (also assuming that the variation of the input data rate is within a limited range). Compared to using the PLL based CLKR whose response can be very slow, the buffer size can be significantly reduced in this case due to the quick response of the TAF-DPS synthesizer.

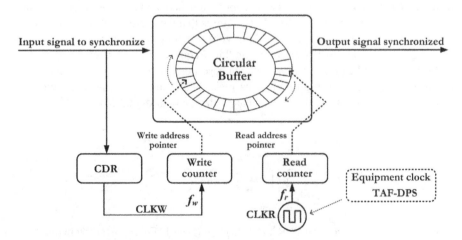

FIGURE 5.9   TAF-DPS enabled buffer architecture inside the receiver.

This problem can also be investigated using the concept of a frequency field. In this application (and the one discussed in Section 5.2), there are two closely related but different parameters: the data rate and the clock frequency. The data rate in the TX is variable or is not explicitly known to the RX. To successfully transfer data between the TX and RX, the data rates in the TX and RX must be equal. In other words, there is only one data rate in the whole system. However, several frequencies could coexist in the system. The various circuit blocks are located in their corresponding frequency fields. One or several such frequency fields are time-average-frequency field. The field is dynamically adjustable so that the data rates on both sides can be made equal. The TAF-DPS, which works in the background, is the source to generate this time-average-frequency field.

5.4 CLOCK DATA RECOVERY IN TIME-AVERAGE-FREQUENCY FIELD

$$T_{txavg} = T_{rxavg} = a_0T_0 + a_1T_1 + a_2T_2 + \cdots + a_nT_n \qquad \sum_{a_i} = 1 \qquad (5.4)$$

In the architecture of Figure 5.8, the clock recovery function is achieved by a circuit commonly called clock data recovery (CDR). CDR plays an important role in digital data communication. At the circuit level, the block diagram of the CDR architecture is illustrated on the left-hand side of Figure 5.10. The goal of this feedback loop is to satisfy equation (5.4), where T_{txavg} $(= 1/f_{txavg})$ is the TX average clock frequency (period) and T_{rxavg} $(= 1/f_{rxavg})$ is the RX average clock frequency (period); T_0, T_1, T_2, ... , T_n are the periods of the types of individual clock cycle in RX; and $a_0, a_1, ... , a_n$ are the possibilities of occurrence of the corresponding cycle types. It is important to point out that, during the clock recovery process, the frequency matching between the TX and RX is achieved in a long-term average sense (over multiple cycles). From each individual cycle's perspective, it is meaningless to talk about frequency matching (since frequency itself is a concept established in the long term of 1 s). The real requirement designated to the CDR circuit is that the average rates of both sides are matched over multiple cycles so that, within each cycle, the incoming data can be reliably latched by the RX circuit using the "frequency-matched" clock signal.

In modern applications, a binary edge/phase detector is often used because of its speed advantage (the high data rates make it difficult for a linear detector to work reliably). The output from the binary detector is in digital fashion. However,

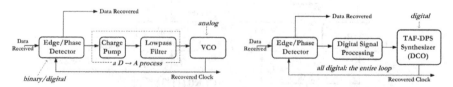

FIGURE 5.10 Conventional CDR architecture (left) and TAF-CDR architecture (right).

as seen in Figure 5.10, the local oscillator (VCO) is controlled by an analog voltage. Therefore, there is a D $\rightarrow$ A process required between the detector and the VCO (charge pump, analog filter, etc). This usually results in high cost (area and power) and slow response. For a typical CDR loop, to make equation (5.4) approximately true, n could be a very large number due to the slow loop response. This has negative impact on the CDR's frequency tracking capability and jitter tolerance.

On the right-hand side of Figure 5.10, the time-average-frequency CDR (TAF-CDR) architecture is depicted. One of the key benefits of TAF-CDR is the elimination of the D $\rightarrow$ A process required in conventional CDR. Furthermore, since the loop has very fast response, the frequency tracking capability and jitter tolerance can be greatly enhanced. In this approach, the enabling block is the TAF-DPS based digital controlled oscillator (DCO), which takes a digital value as its frequency control variable. As can be seen, TAF-CDR takes advantage of the detector's digital output. Consequently, the entire loop is digital. The loop latency M, which is the time elapsed (in number of RX cycles) between the *moment of detecting* and the *moment of action*, can be precisely determined.

Instead of using many frequencies (periods) to match the two sides, as indicated in (5.4), the DCO only generates several discrete frequencies (periods) and uses them to achieve the matching goal. For example, three unique cycles (or frequencies or periods) can be produced to serve the three purposes of *hold*, *speed up*, and *slow down*, respectively. In the short term (within M cycles), the loop dynamically matches TX and RX frequencies, as shown in (5.5). Although the frequency matching is only an approximation in the short term, it is good to the degree that the TX data can be reliably captured by the RX. From the perspective of each data unit, the setup-and-hold margin of the sampling cell inside the RX can be guaranteed by satisfying (5.5). In the long term, (5.4) is satisfied as well just as in the case of conventional CDR. As a result, no data are lost or created during this frequency matching process. Using Flying-Adder as an exemplary implementation, Section 6.8 of [Xiu12a] has an extensive discussion on the mechanism of TAF-CDR:

$$T_{\text{txavg}} \approx T_{\text{rxavg}} = \sum_{i=1}^{M} \alpha_i T_i \qquad \sum_{i=1}^{M} \alpha_i = 1 \qquad (5.5)$$

Clock data recovery can be more conveniently explained using the concept of a frequency field. In this case, there is only one data rate and it is determined by the TX circuit. The data rate in the RX block must be made equal to this rate. The RX circuit sits in a frequency field whose frequency value is constantly changing [refer to equation (5.4)]. In the TAF-CDR case, this frequency field only has three frequency values. The selection of the value is made in every M cycles by a calculation block [refer to equation (5.5)]. The producing source of this frequency field is the DCO (TAF-DPS). In the following paragraphs, two examples, $M = 3$ and $M = 1$, are used to explain this CDR operating in a time-average-frequency field in further detail.

Figure 5.11 shows a transistor-level simulation result for a data rate of about 300 Mbps. The design is implemented in a 180-nm CMOS process. The loop latency

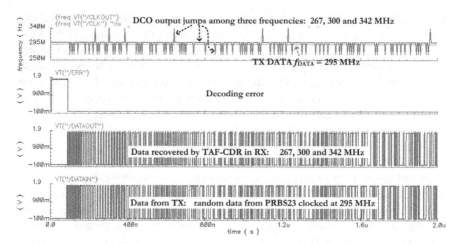

FIGURE 5.11   Simulation of a TAF-CDR operating around 300 MHz using $M = 3$.

is designed as $M = 3$. In this simulation, the incoming data (from TX) are intentionally driven by a clock of 295 MHz (0.16% away from the normal frequency). The DCO in the RX is designed to generate three discrete frequencies: 300 (hold), 342 (speed up), and 267 MHz (slow down). These three frequencies are produced by a TAF-DPS using the configurations 8Δ, 7Δ, and 9Δ, respectively. The base time unit Δ = 417 ps is generated from a four-differential-stage, eight-output CMOS ring oscillator that runs at 300 MHz by locking to a 15-MHz reference.

As shown in the top trace of the DCO output frequency measurement over time (more precisely, it is the measurement of each pulse length), the TAF-CDR dynamically adjusts its DCO output frequency every three cycles. At any particular moment, it selects one frequency among the three choices of 267, 300, and 342 MHz. The decision is made based on the relationship between the edges of the incoming data and the RX clock. It is made in real time within the detector and the digital signal processing blocks. It is outputted to the DCO to produce a corresponding action. The bottom trace is the incoming TX data, which are generated from a 23-bit PRBS (pseudorandom binary sequence) encoder driven by the 295-MHz clock. The second trace from the bottom is the recovered data (outputted from the detector). The third trace is the output from an error detector which is made of the corresponding 23-bit PRBS decoder. The inherent logics of the PRBS encoder and decoder are matched. The error detector is fed by the recovered data (the second trace) and is driven by the recovered clock (the DCO output). Thus, it will report error (logic high) if the recovered data contain error. The simulation shows that the incoming data are correctly latched by the receiver. It also indicates that the clock is rightfully recovered/extracted from the incoming data.

This simulation shows that this TAF-CDR circuit has correctly recovered the transmitted data under the condition that the incoming data rate (295 MHz) is 16,666 ppm away from its normal 300 MHz. The top trace clearly shows that the TX data rate

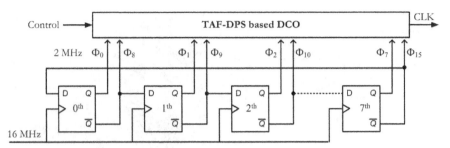

FIGURE 5.12   Digital ring of 8 D-type flip-flops and digital DCO.

of 295 MHz is averagely-matched by the DCO output. As expected, more samples of slowdown frequency (267 MHz) are used than of speedup frequency (342 MHz) since incoming data are slower than the normal value of 300 MHz. However, the majority of the samples are at 300 MHz (which is the hold state). In the construction of this system, the design of the DCO circuit is not the focus; the dynamic field it creates is the essence of this architecture.

As shown on the right-hand side of Figure 5.10, the architecture of TAF-CDR is straightforward. This can lead to elegant and extremely low cost implementation in some cases. One such example is an ultralow-power and low-cost CDR function ($<1\mu$W) for a low-data-rate application of 2 Mbps. The goal of low cost is achieved by a TAF-CDR circuit which is entirely made of standard cells. This design is implemented in a 65 nm CMOS process. The input of the DCO is made of a ring of 8 D-type flip-flops which are driven by a 16 MHz reference signal, as illustrated in Figure 5.12. The resulting 16 outputs are in 2 MHz frequency and are fed to the TAF-DPS based DCO. The DCO is also made of digital standard cells. The control signal *fed* to the DCO comes from the detector output (after it is processed by the digital signal process block; refer to the right side of Figure 5.10).

The base unit $\Delta = 31.25$ ns is generated from the 16 outputs of 2 MHz ($\Delta = 500$ ns/16). The DCO is responsible for generating three types of cycles of 15Δ (468.75 ns), 16Δ (500 ns), and 17Δ (531.25 ns). Each cycle is used in the appropriate time so that the incoming data flow can be tracked. Since the bit time is around 500 ns ($\sim$2 Mpbs) in this case, the operation of detect–calculate–apply in the TAF-CDR loop (refer to the right side of Figure 5.10.) can be finished within one cycle. Thus, the latency is designed as $M = 1$ (in contrast to $M = 3$ in the previous case where bit time is 3.3 ns and it operates in a slower process of 180 nm). As a result of this minimum latency, this TAF-CDR circuit has a very fast response speed. This leads to a large frequency error (and jitter) tolerance.

Figure 5.13 shows the test configuration used in the laboratory to evaluate this TAF-CDR. An Agilent E4437B signal generator is used to provide the test data stream. Its internal BER tester is used to check the received data stream. As shown, a 2 Mbps stream of PRBS data is sent from the signal generator to the chip under test. The recovered data from the TAF-CDR is sent back to the signal generator for validation. The 16 MHz reference needed for the DCO is provided by a function

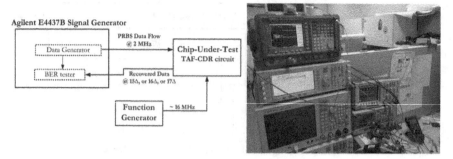

FIGURE 5.13   Test configuration of TAF-CDR.

generator. In this way, the frequency of this reference can be adjusted so that the frequency tolerance of this clock data recovery system can be studied.

Figure 5.14 shows some of the experimental result. The snapshot on the right-hand side is the BER test result. The input data are a 2 Mbps data stream of PN9. The number of bits used is 4,294,967,295 ($=2^{32} - 1$). The BER is zero. The screen snapshot on the left illustrates the waveforms of incoming data (top), recovered data (middle), and recovered clock (bottom). As seen, the recovered data follow the incoming data. However, the data are delayed by one cycle because of the latency $M = 1$. From the recovered clock waveform, if inspected closely, it can be seen that there are three types of cycles of different lengths. It is the evidence of the slowdown, hold and speedup cycles. For example, the cycle marked by the markers is measured as 540 ns, which is the slowdown cycle. This fact is more clearly revealed in Figure 5.15.

The oscilloscope snapshot on the left side of Figure 5.15 is obtained in the single-run mode. The bottom waveform is the incoming data of pattern 0001. The top waveform is the recovered data, which show the pattern 1110 (there is a level of signal inversion added in this case). The waveform in the middle is the recovered clock outputted from the TAF-CDR. If viewed closely, it can be seen that the input data are always at 500 ns length. But the recovered data and clock have several different lengths. This fact is easier to be viewed in the snapshot on the right-hand

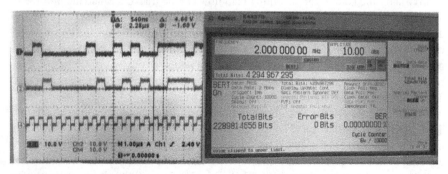

FIGURE 5.14   Experimental result of ultralow-power TAF-CDR operating at around 2 MHz using $M = 1$: waveforms (left) and BER result (right).

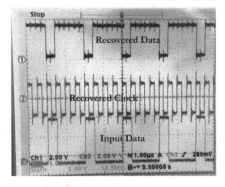 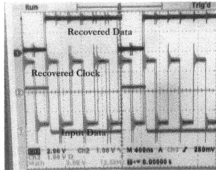

FIGURE 5.15   Illustration of three types of cycles used in TAF-CDR: single run (left) and continuously triggered (right).

side, which is continuously triggered. These data are another evidence to support the statement that a "jittery" clock is not necessarily a bad thing (refer to Section 3.6 and Figure 3.11).

Due to the small latency $M = 1$, this TAF-CDR circuit has a large tolerance for frequency error. The DCO normal frequency (the frequency in the hold state) is usually designed to be equal to the normal rate of the transmitted data stream (2 MHz in this case). When the incoming data rate deviates from the normal value, the TAF-CDR needs to track it. This feature is tested in the laboratory by using the function generator as a adjustable frequency source. During the test, the frequency of the 16 MHz reference is adjusted (this is equivalent to adjusting the incoming data rate) and the BER is checked in each case. Tables 5.1 and 5.2 are the results of using data streams PN9 and PN15, respectively. In all cases, the number of bits used is 214,748,3647 ($=2^{31} - 1$). The frequency has been adjusted from both directions: frequency+ and frequency−. It is worth pointing out that this TAF-CDR circuit is designed for the PN5 data stream (since 8b/10b encoding is used and consequently

TABLE 5.1   BER at Various Degrees of Frequency Error Using Data Stream PN9

Error	Frequency+ (MHz)	BER+	Frequency− (MHz)	BER−
1 ppm	16.000016	0	15.999984	0
10 ppm	16.00016	0	15.99984	0
100 ppm	16.0016	0.0000001	15.9984	0.0000002
500 ppm	16.008	0.000001	15.992	0.000001
1000 ppm	16.016	0.00002	15.984	0.00003
5000 ppm	16.08	0.000135	15.92	0.00009
1%	16.16	0.0002	15.84	0.0001
2%	16.32	0.0009	15.68	0.001
3%	16.48	0.0025	15.52	0.0034
4%	16.64	0.0085	15.36	0.0093
5%	16.8	0.012	15.2	0.0115

TABLE 5.2 BER at Various Degrees of Frequency Error Using Data Stream PN15

Error	Frequency+ (MHz)	BER+	Frequency− (MHz)	BER−
1 ppm	16.000016	0.00000001	15.999984	0.00000002
10 ppm	16.00016	0.0000007	15.99984	0.00000065
100 ppm	16.0016	0.0000087	15.9984	0.00003
500 ppm	16.008	0.00008	15.992	0.00008
1000 ppm	16.016	0.0007	15.984	0.00062
5000 ppm	16.08	0.0012	15.92	0.0009
1%	16.16	0.0019	15.84	0.0012
2%	16.32	0.0028	15.68	0.0019
3%	16.48	0.0054	15.52	0.0047
4%	16.64	0.0174	15.36	0.0185
5%	16.8	0.022	15.2	0.031

the longest no-transition data stream is 5 bits). Better results than that of PN9 and PN15 can be expected since more transition activities are available in PN5. It is also interesting to mention that, if the use of different pulse lengths is preferred to be regarded as jitter (by people of old fashion), then the *TAF-CDR circuit treats deterministic jitter with response speed.*

5.5 NETWORK-ON-CHIP GALS STRATEGY IN TIME-AVERAGE-FREQUENCY FIELD

With billions of transistors used in today's large chips, the advantage of uniprocessor architectures is diminishing due to its demand for high power, high clock frequency, and the global distribution of clock signals. Multicore chips are emerging as the prevailing architecture in both general-purpose and application-specific markets since this architecture allows the distribution of the computation load to multiple cores which can operate at their optimum speeds (clock frequencies). Consequently, *the challenge in architecture design is shifted from computation to communication.* As the core count increases, the need for a scalable on-chip communication architecture that can deliver high bandwidth becomes a necessity. Traditionally, the bus has been the dominant structure for SoC communication, as shown on the left in Figure 5.16. However, it does not scale well with the increased number of cores. This leads to the recent architecture of networked on-chip communication [Bje06, Kol09, Val10]. In the middle drawing of Figure 5.16, every computing module is surrounded by communication links and routers. From any source to any destination, data are routed by logical or physical links using a predefined protocol. The drawing at the right shows a generic network-on-chip (NoC) architecture where multiple cores are interconnected with sophisticated on-chip networks. Each computing module is supported by a pair of interface adapter and routing module. All are connected on-chip in a meshed network.

A key characteristic of the NoC is heterogeneous clocking. Within a SoC designed with NoC methodology, each computing module can operate at its own optimum

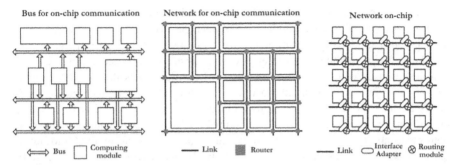

FIGURE 5.16 Bus architecture for on-chip communication (left), network architecture for on-chip communication (middle), and generic NoC architecture (right).

clock frequency. As a result, power consumption can be optimized based on the loading of each module. More importantly, the large amount of power consumed by the high-frequency global clock distribution network is eliminated. NoC architecture also alleviates the signal delay problem associated with the time of flight of long wires. The NoC is a SoC design strategy that separates the tasks of computation and communication in a controlled way so that each can be addressed efficiently. It also facilitates the integration of various IPs, which could come from different vendors for a particular SoC. A direct consequence of the NoC is the existence of many frequency fields in a SoC, as illustrated on the left in Figure 5.17 where there are 25 cores (or IP blocks) and each has its unique operating frequency.

In the NoC architecture, a key design challenge (as a result of heterogeneous clocking) is the synchronization among the cores that may run synchronously at different frequencies or operate at a mixed synchronous–asynchronous mode [Che04, Mul05]. For each computing module (or other type of IP), implementation can be synchronous, which is a common and efficient design method used by today's

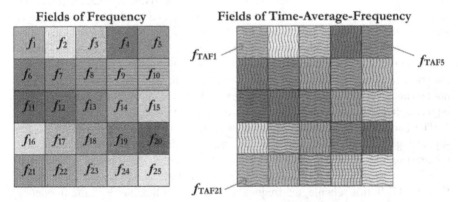

FIGURE 5.17 Existence of many frequency fields in a NoC (left) and flexible time-average-frequency fields for NoC (right).

designer. Communication among modules is carried out in the asynchronous style to accommodate the difference in operating frequencies. This is the so-called globally asynchronous and locally synchronous (GALS) strategy [Mul05, Tee07]. When data transfer between modules of different frequencies needs to be accomplished, there is a design issue of preventing metastability. Metastability is the condition that a signal's voltage level is at an intermediate level: It can be interpreted neither as a 0 nor as 1. This condition may persist for an indeterminate amount of time (refer to the discussion presented in Section 2.5 and Figure 2.5). Two methods can be used to deal with metastability: timing-safe and value-safe. The timing-safe method allocates a fixed period of time for metastability to be resolved (e.g., two flip-flops synchronized). The value-safe method waits for metastability to be resolved by stretching or pausing the clock signal (refer to Section 4.2).

Figure 5.18 depicts two circuit techniques which can be included in the interface adapter in the NoC architecture of Figure 5.16 to handle asynchronous data transfer. A FIFO module resides between the transmitter (TX) and the receiver (RX). It has ports DIN, OUT, and OK_to_PUT to face the TX and ports DOUT, TAKE, and OK_to_TAKE to face the RX. In the drawing on the left, the TX clock CLKT and RX clock CLKR are generated locally. They can be paused (stopped) or stretched based on the FIFO status. This circuit employs the value-safe approach. In the drawing on the right, CLKT and CLKR are provided from the TX and RX, respectively. They are not locally controllable (e.g., they cannot be paused or stretched by the local circuit surrounding the FIFO). The OK_to_PUT and OK_to_TAKE signals from the FIFO are synchronized by the synchronizers to the CLKT and CLKR domains, respectively. This circuit utilizes the timing-safe method and two clock cycles are used to resolve metastability.

In the GALS design strategy, local modules are synchronous blocks that can be designed and implemented using conventional CAD tools with which most designers are familiar. While interfacing the modules, however, special care needs to be given so that metastability can be avoided for the various interfacing scenarios: the same frequency but different phase (mesochronous), the same frequency on average but not exactly the same (plesiochronous), and not the same frequency (heterochronous). The circuit techniques shown in Figure 5.18 were invented to address this challenge. However, they are not without problems. The value-safe approach (on the left) relies on the ring oscillator, which must be stopped and started frequently. This can significantly degrade the clock quality (large jitter). The timing-save method (on the right) uses only two (or three) cycles to allow the metastability to settle down, which might not be enough if higher reliability is required. Moreover, this method has the cost of larger latency due to the multicycle synchronizers used.

The two types of circuits described in Figure 5.18. illustrate the interface for supporting communication between the frequency fields of Figure 5.17. (the left scheme). Using the flexibility provided by the time-average-frequency field, it is possible to build a more cost efficient NoC structure, as shown on the right in Figure 5.17. In this scheme, the frequency fields are all flexible (e.g., the frequency value of each field is constantly changing with small granularity) so that data flows among them can be better managed. The architecture in Figure 5.19 uses TAF-DPS to create this time-average-frequency field in each core and to address the GALS

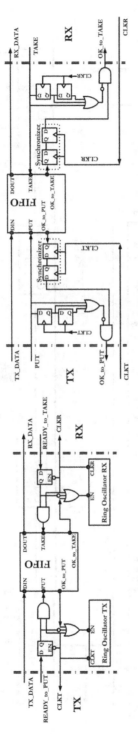

FIGURE 5.18 Interface for GALS: stretching or pausing clock (left) and using synchronizer (right).

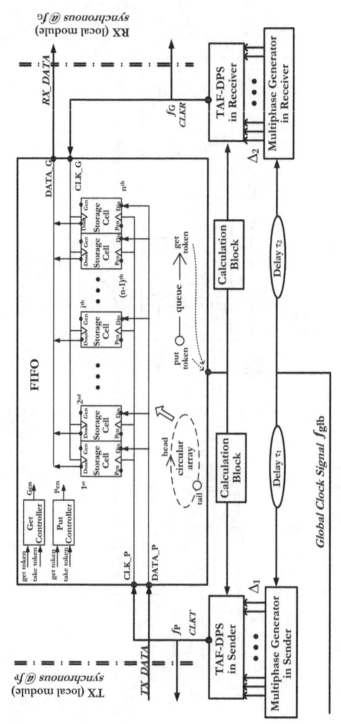

FIGURE 5.19 Architecture using TAF-DPS for interface in GALS.

interfacing challenge. A FIFO with n storage cells resides between two interfacing modules that are synchronously operating at f_P and f_G, respectively. Each storage cell has separated write and read ports that are controlled by write and read clocks independently. The write-and-read operations are enabled by their respective enable signals. All cells' data ports are connected to the common data buses DATA_P and DATA_G. At any given moment, only one cell is selected for read or write. Data in this FIFO are immobile. In other words, once data are latched into the FIFO, it will not be moved between cells. All the storage cells form a circular array. Two tokens control the input and output behavior of the FIFO. Put_token is used to enqueue data items and take_token is for dequeuing data items. Put_token is at the tail and take_token is at the head. Once a token is used by a cell for a data operation, it is passed to the next cell. This FIFO design is similar to the synchronous–synchronous FIFO in [Che04].

The two FIFO clocks CLK_P and CLK_G are produced by two TAF-DPS synthesizers that are configured to generate f_P and f_G, respectively. The base units used by the synthesizers, Δ_1 and Δ_2, are generated from the multiphase generators in the sender and receiver. The reference for the multiphase generators is the global clock signal at frequency f_{glb}. For the purposes of low-power operation and high-quality distribution at long distances, f_{glb} is preferably chosen at a low value (such as below 100 MHz). Ideally, the circuit for the multiphase generator is a PLL or DLL. The TAF-DPS can generate higher frequencies (refer to Figure 3.5) for CLK_P and CLK_G. Based on the FIFO status, both the CLK_P and CLK_G can be stopped and stretched, as shown in the left and middle simulation results in Figure 5.20. The head and tail of the queue can be properly adjusted in real time through the clock-stop and clock-stretch based on the put and get token fed to the TAF-DPS synthesizers. In certain cases, the CLK_G's frequency f_G can be adjusted dynamically according to the sender's data rate if it is known or can be detected. This scenario is shown in the simulation on the right-hand side of Figure 5.20.

Compared to the existing circuit techniques used in the GALS synchronization interfacing, the advantages of this TAF-DPS architecture are as follows:

- It eliminates handshake signals.
- The FIFO size is reduced (this is made possible by the TAF-DPS's fast response).
- Lower power consumption is possible (this is due to the data immobile inside the FIFO).
- There is no latency penalty (since no multicycle synchronizer is used).
- Low-frequency global clock distribution (this is enabled by the TAP-DPS's role of functioning as a frequency multiplier).
- One circuit block works for all modes. It can support the mesochronous, plesiochronous, and heterochronous modes without circuit modification. Switching among the operations of the stoppable clock, stretchable clock, and data-driven clock can be carried out dynamically in real time.
- The operating frequency for each core can be optimized and adjusted in real time.
- This interface module can be standardized as a universal IP for GALS.

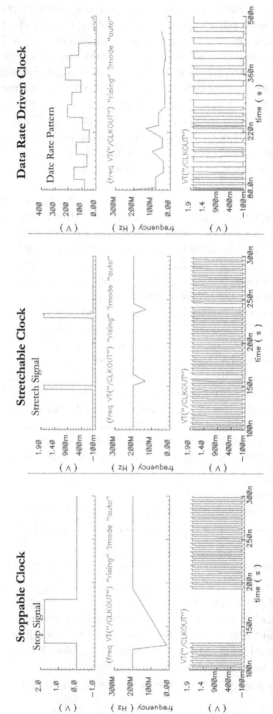

FIGURE 5.20 TAP-DPS interfacing circuit: stoppable clock (left), stretchable clock (right), and data-rate-driven clock (right).

In Figure 5.19, the multiphase generators inside the sender and receiver can be associated with the corresponding computation core (physically in the same module). Their reference signal is the global clock signal at f_{glb}. When reaching each core, this reference signal can either be used directly or be frequency boosted by a PLL. Excluding these multiphase generators, all the rest circuits in Figure 5.19. can be grouped as a standard IP (a physical module). The NoC structure illustrated on the left in Figure 5.17 can be conveniently created by abutting one core (with its associated multiphase generator) to another, with an aforementioned IP in between.

5.6   NETWORK TIME SYNCHRONIZATION IN TIME-AVERAGE-FREQUENCY FIELD: IMPROVED TIME GRANULARITY AND FREQUENCY GRANULARITY

Sections 5.2–5.5 dealt with data communication problems in a local area where there are direct links for clock information to be passed among communicating parties. This scenario occurs most often among blocks within a chip or among systems networked in a small local area network. For problem of this nature, *frequency synchronization* among the clock sources is the target. For large-packet-oriented networks (such as the Internet), a direct clock link (more precisely, direct frequency link) is impossible. In this case *time synchronization*, instead of frequency synchronization, is the target. It has to be achieved through *synchronization message exchange*. In a time division multiple access (TDMA) network, synchronization messages are exchanged in guaranteed time slots. In packet networks, messages are exchanged as regular communication packets [Fer13]. Time synchronization depends on the statistical characteristic of the network (such as network delay). Based on the cost and the required synchronization target, the task of synchronization can be carried out in software, in hardware, or in a hybrid mode. In all these methods, the fundamental building blocks of the time synchronization mechanism are the synchronization event detection technique, remote clock estimation technique, and local clock correction technique. They impact the achievable synchronization precision (and accuracy when synchronizing to an external time reference):

$$\pi = c_1 \epsilon + c_2 P + c_3 G + c_4 u + c_5 G_s \tag{5.6}$$

The synchronization precision is influenced by the parameters shown in (5.6), where π is the precision, ϵ is the transmission delay uncertainty when reading the remote clock, P is the clock drift (due to local oscillator frequency drift), G is the clock reading granularity, u is the rate adjustment granularity, G_S is the clock setting granularity, and $c_{1,2,3,4,5}$ are the weighing factors [Hor04]. The granularities G, G_S, and u are all related to the size of the tick directly or implicitly (the oscillator frequency f_{osc}; refer to the right-hand side of Figure 1.13). The granularities' impact on precision is studied in [Sch03]. The higher the f_{osc} is, the better the precision that can be achieved. This is because the counter within the real-time clock increases in steps of a tick ($=1/f_{osc}$). In all current hardware real-time clock implementations, the time granularity is $1/f_{osc}$ (in unit of seconds). The frequency granularity is also $1/f_{osc}$ (a unitless number). This is because, at finest resolution, one pulse (one tick) can be deleted from a pulse train comprising f_{osc} number of pulses within the time frame of 1 s. If TAF-DPS is used in the real-time clock generator, the time granularity becomes Δ. This is much smaller than $1/f_{osc}$ since $\Delta = 1/(Kf_{osc})$ in most TAF-DPS implementations. At the same time, the frequency granularity becomes $1/(Kf_{osc} - 1)$. This is because, at its finest resolution, the TAF-DPS can remove one Δ from a pulse train of f_{osc} pulses (instead of deleting one entire pulse as in the previous case).

The crystal oscillator (XO), temperature-compensated crystal oscillator (TCXO), and oven-controlled crystal oscillator (OCXO) are the most often used frequency references for today's electronic systems. The OCXO has the best performance on frequency stability at the expense of high cost and large power consumption. For all these oscillators, the best performance is usually achieved at the frequency range of 10–20 MHz [Vig01]. Thus, time granularity is 50 ns and frequency granularity is 5×10^{-8} for a 20 MHz oscillator. Figure 5.21 shows a plan using TAF-DPS to improve the time and frequency granularity. In the block diagram on the left, a same-rate PLL ($N = 1$) is used to generate a plurality of K output signals from its VCO. The value of the base time unit Δ is the time span between any two adjacent outputs and, thus, $\Delta = 1/(Kf_r)$. With today's advanced CMOS technology, Δ can be made very small (large K, more stages), such as $\sim$50 ps ($K = 1024$, $\Delta = 48.8$ ps, 20 MHz oscillator). The time and frequency granularities in this case are 48.8 ps and 4.9×10^{-11}, respectively. Since the PLL operates at the same rate as the reference, the VCO phase noise is not multiplied. The additional jitter added by the TAF-DPS is deterministic and controllable [Xiu12b]. Overall, the *time* outputted from this block has significantly improved granularity, which can help improve the time synchronization precision as can be understood from (5.6). The block diagram on the right shows the DLL-based approach. Compared to the PLL-based design, the DLL method eliminates the noise contribution from the VCO. The key point from the designs illustrated in this figure is the intention of taking advantage of the fine time resolution, which is made possible by the fast transistor available with today's advanced CMOS technology. Those blocks can be conveniently integrated in the main processing chip to reduce the overall system cost. Table 5.3 lists two examples of improvement achieved in time and frequency granularities using $K = 128$ and $K = 1024$.

Equation (5.6) indicates that oscillator frequency drift is also a major factor in limiting the achievable synchronization precision. In operation, frequency drift is mainly induced by temperature change and component aging. The frequency drift contributions from both effects have been studied intensively and can be captured for each oscillator used in a system. The result can be stored in the system to support calibration tables. Based on the data stored in these tables, TAF-DPS can be utilized to easily compensate for the frequency drift since the TAF-DPS frequency transfer function is fixed and known to the designer (linear in a small range around any frequency point). Because of this easy incorporation of the temperature compensation in a system, a cost reduction benefit is the potential elimination of the use of TXCO and OCXO. Those temperature and aging compensations assisted by TAF-DPS can help achieve better network time synchronization with lower cost and lower power consumption.

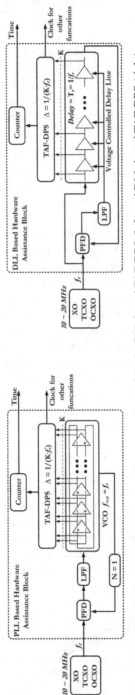

FIGURE 5.21 Improving time and frequency granularities by PLL-based TAF-DPS (left) and DLL-based TAF-DPS (right).

TABLE 5.3   Time and Frequency Granularities Improved by TAF-DPS

f_{osc}	Time Granularity	Frequency Granularity	$K = 128$ Time Granularity	$K = 128$ Frequency Granularity	$K = 1024$ Time Granularity	$K = 1024$ Frequency Granularity
1 MHz	1 μs	10^{-6}	7.8 ns	7.8×10^{-9}	977 ps	9.8×10^{-10}
10 MHz	100 ns	10^{-7}	781 ps	7.8×10^{-10}	97.7 ps	9.8×10^{-11}
20 MHz	50 ns	5×10^{-8}	390 ps	3.9×10^{-10}	48.8 ps	4.9×10^{-11}

Many bus systems for safety-critical embedded systems are time triggered, employing a TDMA protocol in general. Each has its own communications protocol where clock synchronization is integrated and well established. Examples include those designed for avionic and automotive systems: MAFT, Multicomputer Architecture for Fault-Tolerance, a distributed system designed to provide reliable computation in launch vehicle avionics systems; SAFEbus, developed by Honeywell to serve as the core of the Boeing 777 information management system that supports several critical functions, such as cockpit displays and airplane data gateways; and MARS, Maintainable Real-Time System, a fault-tolerant distributed system for process control. These fault-tolerant real-time control systems have stringent requirements on clock synchronization. Further, these systems are usually very cost sensitive since they target the consumer electronic market. In these applications, TAF-DPS can function as an enabling technology for cheap–better–faster products.

Figure 5.22 shows The FlexRay bus system developed by a consortium including BMW, DaimlerChrysler, Motorola, and Philips. It is intended for power train and chassis control in cars [Nat09]. In this system, each ECU (electronic control unit) has its independent clock generator (oscillator). The maximum allowed frequency difference among them is $\pm 0.3\%$. This is a perfect place where TAF-DPS can play the role of improving performance with minimum cost. Frequency error and frequency draft are common challenges that need to be dealt with in designing today's networked equipments. The "*gapped clock*" technique discussed in Chapter 4 is used to cope with this problem. As explained in Chapter 4, TAF-DPS's T_A and T_B interleaved use can be viewed as a much smoother version of the gapped clock. To achieve a certain frequency, instead of removing clock pulse(s) from time to time, TAF-DPS takes a less painful action: Make some cycles slightly longer. As a result, the following cleanup PLL (required by the gapped clock technique) is not needed. This is another example of using TAF-DPS in network time synchronization.

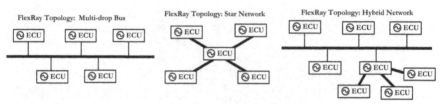

FIGURE 5.22   FlexRay™ communication system architectures: multidrop (left), star (middle), and hybrid (right).

In summary, the task of network time synchronization can be better handled using the time-average-frequency field. Since this field is flexible, synchronization quality can be improved with finer time and frequency granularities.

5.7 CRYSTALLESS REFERENCE AND TIME-AVERAGE-FREQUENCY DRIVEN VLSI SYSTEM: SOURCE FOR FREQUENCY REFERENCE

A trend in modern VLSI system design is to replace the crystal frequency reference with a silicon-based reference [Ken08, Ngu99, Sun06, Mcc09, Sin12]. The silicon frequency source (crystalless) has the advantages of lower cost, higher reliability, easy integration, small form factor, etc. Compared to the crystal-based oscillator, however, the drawback of a silicon source is undeniable: inferior frequency accuracy and stability. This is due to its small Q factor. The left drawing in Figure 5.23 illustrates the generic architecture of a crystalless frequency source. The source of oscillation could be a microelectromechanical system (MEMS) resonator, LCO (*LC* tank-based oscillator), or RCO (*RC*-based oscillator). To bring the oscillation frequency into the desired range, frequency trimming is needed. To counteract the frequency variation caused by environmental change, temperature compensation and supply voltage tracking are required, as shown in the drawing. Both trimming and compensation rely on a frequency tuning mechanism to make the output frequency accurate and stable. The tuning mechanism is oscillator dependent; it is unique for each case of MEMS, LCO, and RCO.

The right-hand side of Figure 5.23 provides an example of a 32-kHz Real Time Clock (RTC) to illustrate the principle of temperature compensation. An oscillator has an oscillation frequency of ∼10 MHz and a frequency variation of ±100 ppm over the temperature range of −40 to 80°C. A frequency–temperature curve can be obtained from the device measurement. From the resulting curve, a corresponding counteracting scheme is stored in a look-up table in the firmware. In operation, the 32-kHz RTC clock is generated by dividing this frequency using a programmable divider. Based on the output from a temperature sensor, the RTC module carries out the counteraction (e.g., adjusts the divide ratio), which makes its output stable at around 32 kHz. This scheme can maintain its output frequency within a smaller variation range [Gri14]. As a result, the frequency stability can be improved about 10-fold (from ±100 ppm to ±10 ppm).

For MEMS, LCO, and RCO, the frequency tuning techniques used for trimming and temperature compensation are all different, but they are all costly in terms of required power and area. Due to its fine frequency granularity and fast frequency switching, TAF-DPS is squarely fitted for this crystalless approach. As demonstrated in Section 3.5, the TAF-DPS frequency resolution can reach the sub-ppm range and the frequency adjustment is dynamically programmable. Based on these features, a scheme is presented in Figure 5.24. From the outputs of temperature and voltage sensors, the TAF-DPS is directed to counteract the frequency variation caused by environmental change. As shown, results from the frequency calibration process (by comparing TAF-DPS output to a known frequency source) and the frequency compensation process (by counteracting environmental influence) can be captured

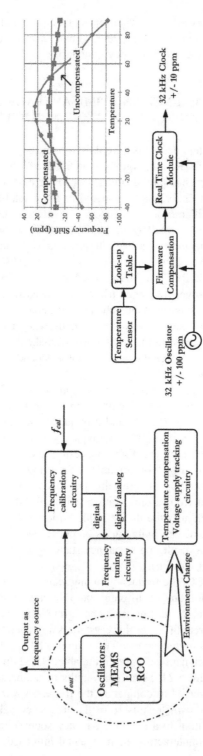

FIGURE 5.23 Architecture of crystalless frequency source with frequency calibration and temperature compensation (left); example of temperature compensation (right).

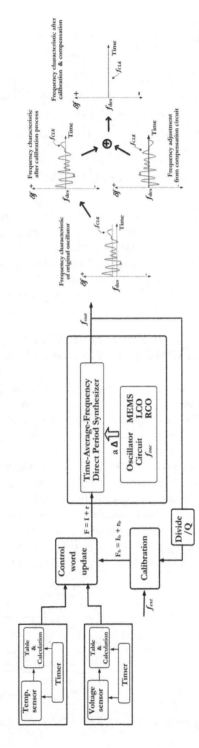

FIGURE 5.24 Architecture using TAF-DPS to improve crystalless frequency source (left); making an unstable and inaccurate frequency source into a stable and accurate one (right).

95

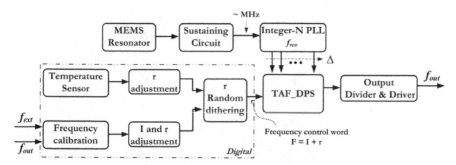

FIGURE 5.25 Architecture using TAF-DPS in MEMS-based frequency source.

into TAF-DPS synthesizer's frequency control word I and r. As a result, an inaccurate and unstable frequency source can be made stable and accurate, as illustrated on the right-hand side of Figure 5.24. This scheme is applicable to all of the MEMS, LCO and RCO. The detail can be found elsewhere [Xiu14a, Xiu14b]. Figures 5.25 and 5.26 are detailed block diagrams of applying this scheme to MEMS and LCO, respectively. Another example using a similar principle is available in [Xiu08].

5.8 CRYSTALLESS REFERENCE AND TIME-AVERAGE-FREQUENCY DRIVEN VLSI SYSTEM: HARMONY IN SYSTEM-LEVEL DESIGN

The major concern in using a silicon-based frequency source, compared to a crystal-based one, is its inferior frequency accuracy and stability. When it is used as the frequency reference source for VLSI systems, chip designers need to (1) make the crystalless frequency source as accurate and stable as possible and (2) make the VLSI system more tolerable to the less accurate and less stable frequency reference source. The first task can be addressed using TAF-DPS as discussed in Section 5.7; the second issue can be addressed by TAF-DPS as well, as will be explained in this section.

In an electronic system, a frequency source with an error of approximately thousands of ppm deviated from its designed value usually does not present a problem to the system if all the circuit blocks are driven by the same clock. However, if multiple systems using different frequency references have data exchange among them, the frequency difference (caused by the said inaccurate references) will be problematic.

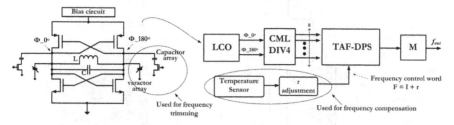

FIGURE 5.26 Architecture using TAF-DPS in LCO-based frequency source.

Large amounts of resources have to be invested to deal with this challenge. This problem, however, presents an opportunity to us for demonstrating the advantages of the time-average-frequency driven VLSI system.

Figure 5.27 illustrates a scenario of two systems that have data communication between them. System A uses a crystalless source A as its frequency reference. From this reference, a PLL or TAF-DPS generates function clock Clock-A at f_{ckA} which is used to drive processor A. The frequency f_{ckA} can deviate from its target value by a certain degree (a few percent in worst case) due to the use of the crystalless source. This fact does not prevent system A's internal operation from being carried out successfully since everything is driven by the same clock. When its data need to be passed to system B, however, this frequency inaccuracy can make the design of system B difficult. To handle the frequency difference between the systems, a CDR circuit must be employed. For a large frequency disparity of several percent, it is impossible for a conventional CDR to handle it (concluded from data published elsewhere and in product datasheets). If system B is designed by the time-average-frequency principle, this problem can be potentially resolved with less painful effort. As shown, a TAF-CDR block can be placed in system B to interface system A. Due to its capability of handing large frequency disparity (refer to Tables 5.1 and 5.2), the TAF-CDR can successfully capture the incoming data even when there is a large degree of frequency difference between the references A and B. In this case, system B operates in the time-average-frequency field. As a result, in terms of function clock Clock-B2 f_{ckB}, system B can adjust itself in fine step and in fast speed.

Therefore, overall at high level, there is an intrinsic harmony in the time-average-frequency driven architecture. At the component level, TAF-DPS can improve crystalless frequency reference's accuracy and stability (Section 5.7). At the application level a system driven by a TAF-DPS clock generator (time-average-frequency system) can tolerate a less accurate and less stable reference source. As a result, the benefit presented by this design approach is significant cost reduction in the bill of materials (BOM) of the electronic system (silicon cost, board cost, power consumption, etc.).

5.9 EFFICIENT IMPLEMENTATION OF MCXO SYSTEM IN TIME-AVERAGE-FREQUENCY FIELD

All frequency sources, whether crystal based or silicon based, suffer from the problem of frequency variation caused by temperature change. TCXO and OCXO utilize an external temperature sensor to sense the environmental temperature around a crystal and direct the compensation circuit to counteract the frequency variation. However, there are challenges of thermal lag and temperature gradients associated with an external temperature sensor. The MCXO (microcomputer-compensated crystal oscillator) system uses self-sensing to do the frequency stabilization. The crystal oscillator in the MCXO is free to vary with temperature, but the fundamental (f_1) and third overtone (f_3) are simultaneously excited. A beat frequency is generated by subtracting the fundamental from the third overtone: $f_b = f_3 - 3f_1$ (or $f_b = f_3/3 - f_1$). This beat frequency is used as the temperature indicator to correct the fundamental or the third

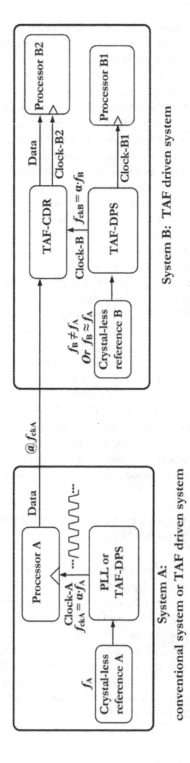

System A:
conventional system or TAF driven system

System B: TAF driven system

FIGURE 5.27 Time-average-frequency driven VLSI system is naturally in harmony with crystalless frequency reference.

overtone, either of which can be used as the MCXO output [Sch89, Fil89, Ben89, Ben91, Blo89, Jac96].

The left-hand side of Figure 5.28 illustrates, the characteristic of frequency vs. temperature (f vs. T) of the fundamental and third overtone for a SC-cut crystal operating in the c mode. The temperature dependences of f_1 and f_3 can be described by (5.7), where a, b, and c are constants, T_0 is a reference temperature, and $\Delta T = T - T_0$. The beat frequency f_b is derived in (5.8). For a temperature range of practical use, the first term is much greater than the higher order terms. Therefore, the beat frequency varies almost linearly with temperature, as described in (5.9) and illustrated in the middle drawing of Figure 5.28. For this reason, the beat frequency of SC-cut crystal can function as a temperature senor which reports the temperature exactly at the crystal's operating point. Therefore, the frequency-vs.-f_b (f-vs.-f_b) curve, illustrated on the right-hand side of Figure 5.28, achieves the same function as f vs. T in reporting the effect of frequency variation on temperature:

$$f_1(T) = f_1(T_0) + a_1 \Delta T + b_1 \Delta T^2 + c_1 \Delta T^3 + \cdots$$
$$f_3(T) = f_3(T_0) + a_3 \Delta T + b_3 \Delta T^2 + c_3 \Delta T^3 + \cdots$$

(5.7)

$$f_b(T) = f_b(T_0) + (3a_1 - a_3)\Delta T + (3b_1 - b_3)\Delta T^2 + (3c_1 - c_3)\Delta T^3 + \cdots \quad (5.8)$$

$$f_b(T) \approx f_b(T_0) + (3a_1 - a_3)\Delta T \quad (5.9)$$

Figure 5.29 depicts two architectures that utilize f vs. f_b for stabilizing the output frequency. In both approaches, the beat frequency f_b is generated by dividing f_3 by 3 and mixing the result with f_1. The sum frequency is removed by a low-pass filter and the difference frequency is outputted as the beat frequency. The f-vs.-f_b data (functioning as f vs. T) is obtained and curve fit (seventh polynomial) for each crystal through calibration and is stored in a nonvolatile memory. In the method shown on the left, f_b is measured by a counter driven by f_3 (or f_1). The result, labeled as N_1, is a number representing the beat frequency (the temperature); N_1 is inputted into a microprocessor. From the prestored f-vs.-f_b curve, the microprocessor generates another number N_2 representing the corresponding correction action; N_2 is fed into a direct digital synthesizer (DDS) that produces a frequency f_d which represents the desired correction. The MCXO output f_{out} is generated from a VCXO (voltage-controlled crystal oscillator). The difference of f_{out} and f_3 is generated by a mixer. Both $f_{out} - f_3$ and f_d are fed to a PFD (phase frequency detector) whose output is used to direct the VCXO through a 1x PLL. As a result, the MCXO output $f_{out} = f_3 + f_d$ is a temperature-compensated frequency.

In the DDS-plus-VCXO method, the SC-cut crystal is made in such way that f_3 is slightly lower than the desired frequency in all temperatures over the interested range. For the method presented on the right-hand side of Figure 5.29, the crystal is made in such way that f_3 is slightly higher than the desired frequency. Based on the current f_b value, the correction determination block sends a signal to the pulse deletion circuit that periodically deletes pulses from the pulse train of the

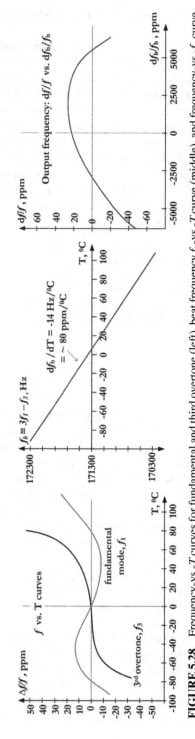

FIGURE 5.28 Frequency-vs.-T curves for fundamental and third overtone (left), beat frequency f_b-vs.-T curve (middle), and frequency-vs.-f_b curve (right).

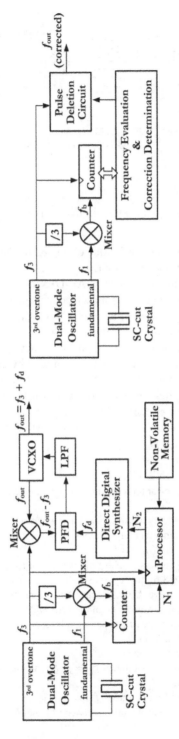

FIGURE 5.29 Two MCXO architectures: DDS plus VCXO (left) and pulse deletion (right).

101

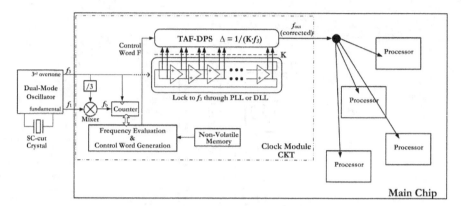

FIGURE 5.30 The architecture block diagram of TAF-DPS MCXO system.

output (similar to the gapped clock technique discussed in Chapter 4), resulting in a temperature-compensated frequency output.

Figure 5.30 describes the method of using TAF-DPS to implement the MCXO system. Similarly, f-vs.-f_b curve is obtained for each crystal and is stored in a nonvolatile memory. The third overtone f_3 is used as a reference to a ring oscillator having K outputs. This ring can be locked to f_3 through either a PLL or a DLL. A TAF-DPS is constructed from these K outputs. Its output frequency $f_{out} = (K/F) f_3$ (refer to Chapter 3). The beat frequency f_b is constantly monitored by a counter. Based on the current f_b value and the prestored f-vs.-f_b table, the control word generation block generates the corresponding frequency control word for the TAF-DPS. As a result, its output is the temperature-compensated frequency.

Figure 5.31 shows the frequency correction scheme. The symbols used in this discussion are defined as follows:

f_{target}	Target frequency for stabilization
f_{out} (T)	MCXO output frequency at temperature T; the goal is to make $f_{out}(T) = f_{target}$ for all temperature points
$f_3(T)$	Third overtone frequency; the raw signal for frequency stabilization (the unstabilized signal); fundamental $f_1(T)$ can be used for this purpose as well
f_{bC}	Beat frequency at which $f_3(f_{bC}) = f_{target}$
C	A point in the f-vs.-f_b curve at which $f_3(T) = f_{target}$ [or $f_3(f_{bC}) = f_{target}$]
O	A point in the TAF-DPS frequency transfer function curve at which $f_{out}(T) = f_{target}$
U	A point in the f-vs.-f_b curve that represents the current temperature; it has a corresponding frequency value $f_3(T)$; in the TAF-DPS frequency transfer function curve, point U represents the uncompensated frequency
df_b	The f_b distance between U and C
$df_3(T)$	The f_3 distance between U and O
F_U	Frequency control word F that makes $f_{out} = f_3(T)$; in other words, no compensation is made
F_O	Frequency control word F that makes $f_{out} = f_{target}$
dF	Distance of frequency control word F between F_U and F_O

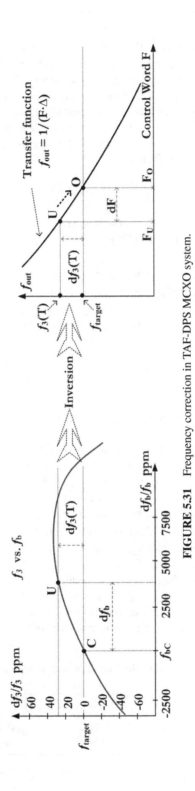

FIGURE 5.31 Frequency correction in TAF-DPS MCXO system.

103

During the calibration process, the point of f_{bC} is indentified and stored. Assume that, at one moment, the counter reports a f_b value that locates at point U (a particular temperature point). From the value of df_b, the frequency offset $df_3(T)$ can be derived from the prestored table. The $df_3(T)$ can be consequently passed to the TAF-DPS for frequency correction. The control word generation block is responsible for producing the dF so that output frequency is moved from point U to point O. The point O is the place where the control word F_O satisfies (5.10). Frequency correction can be described using the steps given below:

$$f_{\text{target}} = 1/(F_O \Delta) \tag{5.10}$$

1. When first powering up, at current temperature T, set the control word $F = F_U = K$ (no correction is made) $\rightarrow$ current output is $f_{\text{out}}(U) = f_3(T)$; most likely, this is not at f_{target}.
2. From the current f_b value, obtain the frequency offset $df_3(T)$ from the prestored table $\rightarrow [f_3(T) - f_{\text{target}}]/f_{\text{target}} = df_3/f_{\text{target}}$ (this is a known value now).
3. Using the relationship $[f_{\text{target}} - f_{\text{out}}(U)]/f_{\text{target}} = [1/(F_O \Delta) - 1/(F_U \Delta)]/[1/(F_O \Delta)]$ $\equiv - [f_3(T) - f_{\text{target}}]/f_{\text{target}} = - df_3/f_{\text{target}}$, calculate the desired frequency control word as

$$F_O = F_U \left(1 + \frac{df_3}{f_{\text{target}}}\right) = K \left(1 + \frac{df_3}{f_{\text{target}}}\right) \tag{5.11}$$

4. Continue to the next cycle of frequency correction.

A numerical example is helpful for understanding the scheme. Assume that $f_3 = {\sim}10$ MHz and a ring of $K = 32$ is used for constructing the TAF-DPS. Initially, the frequency control word F takes the value of $F_U = K = 32$; its output will be the uncompensated crystal frequency $f_3(T)$. Also assume that at current temperature T this frequency deviates from f_{target} by $+10$ ppb ($df_3/f_{\text{target}} = 10^{-8}$). From (5.11), the required F_O can be calculated as $32(1+10^{-8}) = 32.000000032$. Note that the crystal output $f_3(T)$ is 10 ppb faster than f_{target}. The resulting F_O is 10 ppb larger than F_U since the TAF-DPS frequency transfer function has the characteristic of $1/x$ ($\rightarrow df/f = - dF/F$). The beauty of this frequency compensation scheme lies in the facts that (1) although the value of Δ varies with $f_3(T)$, it is not needed in the actual calculation, and (2) the compensation equation is straightforward, as shown in (5.11). Table 5.4 lists the number of bits required in the fractional part of the frequency control word

TABLE 5.4 Number of Fractional Bits Required to Achieve Various Resolutions ($f_3 = {\sim}10$ MHz)

K	$\Delta(f_3)$	$df/f = 1$ ppb	$df/f = 10$ ppb	$df/f = 100$ ppb	$df/f = 1$ ppm
32	${\sim}3.125$ ns	25 bits	22 bits	18 bits	15 bits
64	${\sim}1.5625$ ns	24 bits	21 bits	17 bits	14 bits

F for the cases $K = 32$ and $K = 64$. The equation used for this calculation is 1 LBS = $(df/f)K$.

Compared to the two previously discussed methods illustrated in Figure 5.29, the advantages of TAF-DPS based MCXO are summarized as follows:

- The constraint on the crystal frequency is relaxed. It can be either higher or lower than the target frequency.
- The system bandwidth is much improved (compared to the DDS-plus-VCXO method) since TAF-DPS has a much faster response speed than that of the low-bandwidth PLL used in controlling the VCXO.
- Compared to the pulse deletion method, the signal quality (spurious tones and waveform irregularity) is much improved (see Section 4.3 regarding the gapped clock discussion).
- Compared to the DDS-plus-VCXO method, the resource required is much reduced. Further, the TAF-DPS MCXO system can be designed as an IP module for easy integration into a large SoC system.
- The resource used in the TAF-DPS MCXO system can be further reduced since the PLL/DLL can be shared with those budgeted for the main chip.

5.10   CPU THROTTLING IN TIME-AVERAGE-FREQUENCY FIELD

One of the two key tasks performed by VLSI system is computation. Computation is accomplished by on-chip processors, e.g., CPU, DSP, and GPU. A processor's power consumption is directly proportional to its operating speed and quadratically increased with the operating voltage as shown in (5.12), where C is total capacitance, V is the supply voltage, and f is the operating frequency. The thermal stress induced by high-power consumption in desktop computing and the limited energy capacity of hand-held mobile devices both demand a technology that can reduce the power consumption at the expense of degraded performance when necessary. Dynamical voltage and frequency scaling (DVFS), also known as CPU throttling (e.g., Intel SpeedStep, AMD Cool'nQuiet and PowerNow!, VIA LongHual), is such a technology that the processor's performance and power consumption can be modified while a system is functioning:

$$P = CV^2f \qquad (5.12)$$

Figure 5.32 shows examples of two commercial processors using DVFS technology to control energy consumption. On the left is the Samsung Exynos 4 SoC's voltage–frequency relationship [Vog13]. On the right is the data for the Intel Pentium M [Int04]. When operating frequency is reduced (transistors switch at slower speed), the supply voltage can be lowered as a consequence. This is warranted by laws of physics. Combined together, from (5.12), adjustment on these two variables leads to a significant reduction in power (and thus heat). Figure 5.33 illustrates the

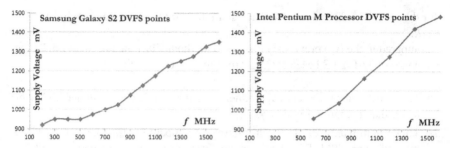

FIGURE 5.32 Voltage–frequency operating points in DVFS technique: Samsung Galaxy S2 (left); Intel Pentium M (right).

working flow of DVFS. Application software detects the CPU status and adjusts the clock generator output frequency through some control register to achieve the desired power–performance trade-off.

The energy (in joules) consumed by an electronic system over a time τ is related to power (in watts or joules per second) is expressed as

$$E(\tau) = \int_0^\tau P(t)\, dt = \int_0^\tau I(t)V(t)dt \qquad (5.13)$$

It is found by researchers that, for a particular computation task, the energy used by a processor has a convex characteristic in plot of energy vs. frequency, as shown in Figure 5.34. [Vog13]. This convexity indicates the fact that there exists an optimal operating frequency in which the energy consumption for the particular task is at the minimal. The energy usage is affected by leakage power and dynamic power consumed by the transistors. Leakage power is independent of the operating frequency

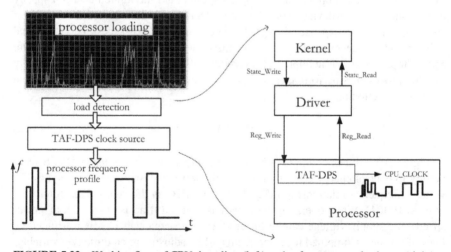

FIGURE 5.33 Working flow of CPU throttling (left) and software control scheme (right).

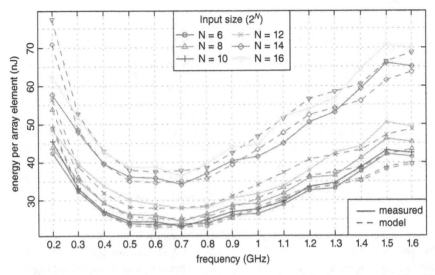

FIGURE 5.34 Existence of an optimum clock frequency for minimal energy usage [Vog13].

but is linearly proportional to the time required to execute the instructions. Dynamic power is influenced by both the operating frequency and the supply voltage. At lower frequency, the time required is extended more than linearly (period is the inversion of frequency) and thus leakage power dominates the power usage trend. This results in a higher energy usage than required in the optimal frequency. At higher frequency, the energy usage is inflated by the quadratic dependence of power on voltage (refer to the fact revealed by Figure 5.32 that supply voltage needs to be raised when frequency is increased). These two opposite trends is a possible explanation of the existence of the optimal operation frequency.

The fast frequency adjustment required by the DVFS and the existence of an optimal operating frequency for minimal energy usage all demand a flexible clock source. The clock generator currently used in DVFS is mostly PLL based, which is hardly flexible. This leads to large processor unavailable time when DVFS is applied. The "clock thinning circuit" used in [Mas05] is not ideal for supporting DVFS either. Neither of them could take the DVFS implementation to the next level. With its flexible frequency generation and frequency switching, the time-average-frequency field is the ideal place for practicing the DVFS techniques. CPU throttling can be carried out in a much more efficient way. The fast frequency switching demonstrated in Figures 3.8, 3.9, and 3.10 can potentially help reduce the processor unavailable time to much below 10 μs (the case of the Intel Pentium 4 [Int04]). Only two clock cycles are required for the TAF-DPS to switch its waveform from one frequency to another. Further, it is achieved seamlessly without the need of pausing the clock during the switching. This is equivalent to 10 ns in the case of 200 MHz, or 1.3 ns for 1.5 GHz. These numbers are much smaller than those used in the PLL based clock generator. In other words, one limiting factor is removed in reducing the processor unavailable time.

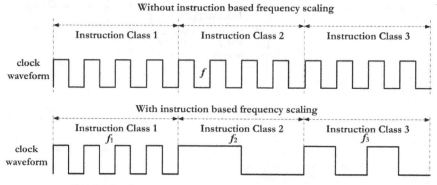

FIGURE 5.35   Instruction-based dynamic frequency scaling.

Another promising approach of DVFS is instruction-based dynamic frequency clocking [Ran98]. This is especially applicable to video/graphical processors where many *add* and *multiply* operations are executed. This approach is motivated from the observation that different functional modules (adder, multiplier, etc.) have different clocking requirements. Hence, based on their critical path delays, they can be driven by different frequencies (more precisely, clock signals of different pulse lengths). This practice asks for "on-the-fly" adjustment of the processor clock speed based on current CPU instruction, as shown in Figure 5.35. For example, in low-level image processing, the "filtering" operation involves mostly add and multiply, "connected component labeling" often requires *comparison*, while "thinning" only needs a simple logic function. Therefore, the same processor can operate at different speeds while performing all these algorithms in the same application environment. This feature of DVFS was not widespread in the past due to the lack of an appropriate clock generator. Cycle by cycle, when operating in a time-average-frequency field enabled by TAF-DPS, the processor can enjoy further power efficiency at the instruction execution level.

In principle, DVFS is in great harmony with the philosophy of time-average-frequency. Also, it is interesting to point out that the dynamical frequency switching used in conventional DVFS implementation does modify the clock's pulse length in real time. This is not different from the pulse length modification in TAF-DPS. They all can be treated the same way within the scope of time-average-frequency although this is not explicitly recognized before.

5.11   SPREADING CLOCK ENERGY IN TIME-AVERAGE-FREQUENCY FIELD

In a clock signal of conventional frequency, only one type of clock cycle exists. Consequently, clock energy is concentrated around one frequency, as illustrated in Figure 5.36. (a clock signal of 888.8 MHz), resulting in strong electromagnetic radiation. In a time-average-frequency field, due to the usage of two or more types of cycles, the clock signal energy is naturally spread. As a tool for spreading a signal

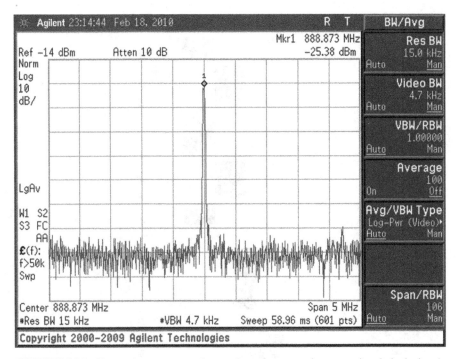

FIGURE 5.36   Energy is concentrated around one frequency in conventional clock signal.

spectrum, TAF-DPS is very efficient and highly precise. The mechanism has been discussed in Sections 5.7 and 6.14 of [Xiu12a]. Figure 5.37 (Figure 6.47 in [Xiu12a]) is reproduced here to illustrate the precision of this method. The spectrum plot on the left is obtained from measurement. It is the spread version of the clock signal in Figure 5.36. The middle plot in Figure 5.37 is the corresponding simulation result displayed using the same scale as in the experiment. As seen, a strong correlation between the two spectra is evidenced. The simulation result on the right reveals some working details that make this scheme of energy spread precise. The top trace is the trend showing the change in the frequency control word. The bottom trace is the TAF-DPS output frequency (period) at each moment of time. The precise control on the output frequency accounts for the highly accurate energy spread.

As stated above, this has been discussed elsewhere [Xiu12a], but this highly accurate spread spectrum method deserves more study, revealing its fine detail from the perspective of time-average-frequency. Figures 5.38–5.40 are associated with an example that serves this purpose. A TAF-DPS is designed using $\Delta = 125$ ps in a 55 nm process. This is achieved from a ring of four-differential-stages running at 1 GHz (by locking to a reference through a PLL). Thus, the output frequency $f_{out} = (K/F)f_{vco} = 8/F$ GHz (or $T_{out} = F \times 125$ ps). The control word F is swept around 8 using the configuration of *magnitude = 0.0625 (100000$_{16}$)* and *step = 0.00067 (002C08$_{16}$)*, as shown on the left in Figure 5.38. The sweep is controlled by a 50 MHz clock (e.g., each step is 20 ns). A full cycle of sweep takes about 7.4, μs. Under this configuration,

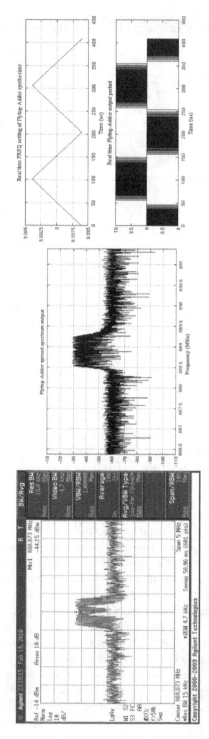

FIGURE 5.37 In a time-average-frequency field, the clock energy spread can be precisely controlled: measurement (left), simulation (middle), and clock cycle distribution (right).

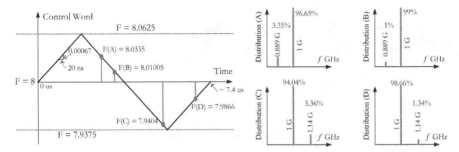

FIGURE 5.38 Spread spectrum achieved by triangular sweep: trend of frequency control word change (left) and frequency distribution of four time segments (right).

the possible TAF-DPS outputs are 0.889 GHz ($T = 9\Delta$), 1 GHz ($T = 8\Delta$), and 1.14 GHz ($T = 7\Delta$). Four segments in the sweep are identified as A, B, C, and D. Their control word values are 8.0335, 8.01005, 7.9404, and 7.9866, respectively. The time span for each segment is 20 ns (it comprises ~20 TAF-DPS cycles). The frequency distributions of the four segments are displayed on the right-hand side of Figure 5.38.

Figure 5.39 includes the experimental data to support the above analysis. The plot on that left shows the frequency distribution; three discrete frequencies of 0.889, 1, and 1.14 GHz are visible. In the middle plot, the frequency-vs.-time trend is displayed in the bottom window. It has a similar pattern to the plot on the right plot of Figure 5.37. Three discrete frequencies are visible. The *frequency density vs. time* varies according to the pattern defined in the control word changing pattern. In the plot on the right-hand side, a zoom-in version of the *frequency density vs. time* is displayed. This frequency range corresponds to the control word range of 7.9375–8. The gradual change of frequency density is clearly visible (from *more-1.14 G-less-1 G* density on the left gradually to *less-1.14 G-more-1 G* on the right). During this process, although only two types of cycles, $T = 7\Delta$ and $T = 8\Delta$, are used, many different frequencies are observed in the spectrum plot since the weight factor (the fraction r) is constantly being updated. This example shows that the time-average-frequency field changes continuously.

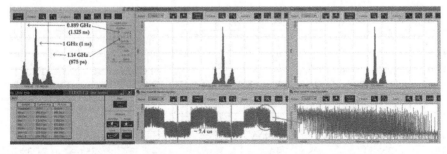

FIGURE 5.39 Frequency distribution (left), frequency-vs.-time plot (middle), and zoom-in frequency-vs.-time plot (right).

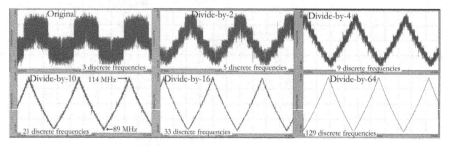

FIGURE 5.40   Measured frequency-vs.-time plots after divider.

As explained in Section 4.6 of [Xiu12a], a divider attached after a TAF-DPS can be viewed as the extension of the TAF-DPS. In other words, the post divider is part of the synthesizer. Figure 5.40 shows the frequency-vs.-time plots of a signal after it is passed through a divider. In previous discussion, using the configuration of Figure 5.38, the TAP-DPS can output three types of cycles: 7Δ, 8Δ, and 9Δ. After passing through a divide-by-2 divider, the signal will have five possible cycles: $(7+7)\Delta = 14\Delta$, $(7+8)\Delta = 15\Delta$, $(8+8)\Delta = 16\Delta$, $(8+9)\Delta = 17\Delta$, and $(9+9)\Delta = 18\Delta$. In general, the number of unique types of cycles after divide-by-M is $N = 2M + 1$. Table 5.5 lists the numbers corresponding to the plots in Figure 5.40. In Figure 5.40, *five frequency-vs.-time* plots are shown, each with a unique divide ratio. It is seen that the numbers of discrete frequencies shown in the experimental plot agree with those in Table 5.5. The output frequency ranges also align with those presented in Table 5.5.

It is also interesting to point out that the divider acts as a low-pass filter that removes all the disturbances associated with the original high frequency. In other words, the divider smoothes the signal. This is clearly evidenced from the plots of divide by 16 and divide by 64. After those divide ratios, the TAF-DPS transfer function characteristic of $1/x$ is gradually visible. Figure 5.41 shows some of the details. On the left is the calculated frequency transfer function plot of F varying from 800Δ to 200Δ using $\Delta = 125$ ps. In the middle is its zoom-in version of F varying from 576Δ to 448Δ. On the right-hand side, the measured data from lab are presented. A high level of correlation is visible.

TABLE 5.5   Types of Cycles After Divider

Divide Ratio	Possible Period T (Frequency), $\Delta = 125$ ps	Number of Unique Types of Cycles	Frequency Range
1	7Δ, 8Δ, 9Δ	3	889 M ↔ 1.14 G
2	14Δ, 15Δ, 16Δ, 17Δ, 18Δ	5	444 M ↔ 570 M
4	28Δ, 29Δ, 30Δ, 31Δ, 32Δ, 33Δ, 34Δ, 35Δ, 36Δ	9	222 M ↔ 285 M
10	70Δ, 71Δ, ..., 80Δ, 81Δ,..., 90Δ	21	88.9 M ↔ 114 M
16	112Δ, 113Δ, ..., 128Δ, 129Δ, ..., 144Δ	33	55.6 M ↔ 71.3 M
64	448Δ, 449Δ, ..., 512Δ, 513Δ, ..., 576Δ	129	13.9 M ↔17.8 M

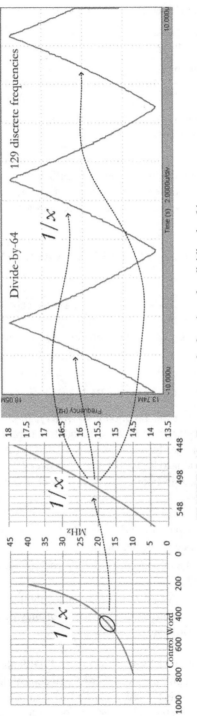

FIGURE 5.41 Frequency transfer function after dividing-by-64.

TABLE 5.6   Comparison between TAF-DPS and Conventional Spread Spectrum Techniques

Jitter Type	Conventional		TAF-DPS	
	No spread	With spread	No spread	With spread
P2P	0	Not directly controllable, hard to predict	Δ	Δ
C2C	0	Finite, varies at low frequency	Δ, fixed pattern	Δ, pattern varies at low frequency

Further, it is worth mentioning that the TAF-DPS spread spectrum does not add additional jitter. Using the concepts of P2P and C2C jitter, the differences between TAF-DPS and conventional spread spectrum techniques are compared in Table 5.6.

5.12   TAF-DPS AS CIRCUIT TECHNIQUE IN BUILDING SPECIAL FUNCTION BLOCKS

From a circuit design perspective, TAF-DPS has the capability of manipulating each individual pulse's length-in-time and duty cycle. This is a useful method for building many circuit blocks, even when the concept of time-average-frequency is not explicitly involved.

5.12.1   TAF-DPS as DCO for PLL and DLL

The TAF-DPS frequency transfer function $f_o = (K/F) f_r$ implies that it can function as a DCO where the frequency word F is the control parameter. In analog VCO, the control input is a voltage. In this case, it is a digital value. This DCO can be used as a frequency generator to build a frequency-locked loop (FLL). Its analogy to VCO-based PLL is depicted in Figure 5.42. The key characteristics of the FLL can be summarized as follows:

- All variables in the loop are digital values; there is no analog voltage. This leads to efficient and accurate implementation.
- The FLL output can be in conventional frequency (F only takes an integer) or in time-average-frequency.
- The FLL loop bandwidth can be programmable, configured by software in real time. This leads to software FLL [only the frequency detector (FD) and DCO are implemented in hardware].

As demonstrated in Section 4.25 of [Xiu12a], TAF-DPS also has phase synthesis capability. In Figure 5.43, a phase control word P can control the time delay between the signals f_o and f_p. Using this feature, a digital PLL can be constructed from TAF-DPS as well. As illustrated on the right-hand side of Figure 5.43, there are two loops

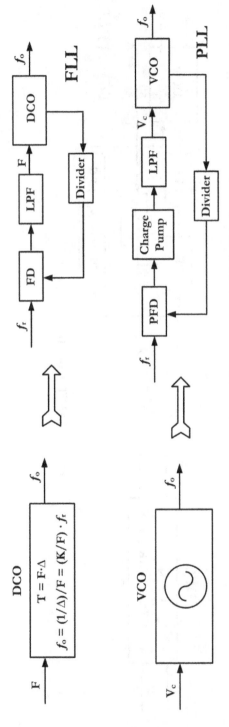

FIGURE 5.42 TAF-DPS DCO for FLL (top) and VCO for PLL (bottom).

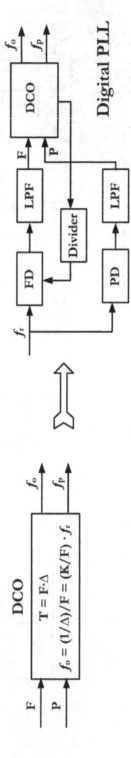

FIGURE 5.43 TAF-DPS DCO for digital PLL.

116

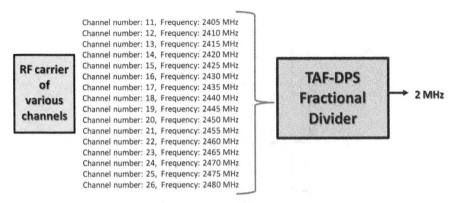

FIGURE 5.44   Application scenario requiring fractional divide ratio (Zigbee).

in this digital PLL: the frequency loop, which is responsible for synthesizing the appropriate frequency; and the phase loop, which generates the desired phase. In this view, it can also be regarded as a DLL.

5.12.2   TAF-DPS as Open-Loop Fractional Divider

Fractional frequency division is a common task in modern IC design. Figure 5.44 is an example of such an application. ZigBee is a wireless person area network (WPAN) that is built upon the physical layer and media access control (MAC) defined in IEEE 802.15.4. Its key implementation requirement is low cost and low power. Figure 5.44 shows the requirement of generating the 2 MHz baseband clock from all the 16 RF channels using the 2.4 GHz industrial, scientific, and medical (ISM) band. Of those 16 channels, half of them have a noninteger ratio between the RF and the baseband frequencies. A fractional PLL can be used to fulfill the requirement. However, the associated cost might be prohibitive for it being used in low-cost and low-power system. From the TAF-DPS frequency transfer function $f_o = (K/F)f_r = (Kf_r)/F$, it is seen that the TAF-DPS can function as a divider of frequency Kf_r with ratio F. The reason is that TAF-DPS can reach the input signal's internal phases, as illustrated in Figure 5.45.

Furthermore, the ratio can be fractional if F is allowed to use a fraction, in which case the resulting signal is a time-average-frequency. However, if a conventional frequency is required in the resultant signal, a post divider can be used to recover the fraction used. This technique of PDFR is explained in Section 4.6 of [Xiu12a]. Using TAF-DPS and PDFR, a scheme to achieve fractional division between a source frequency and a destination frequency is illustrated in Figure 5.46. The N divider is used to generate multiple phases from the source signal. The TAF-DPS performs the fractional division. The purpose of the M divider is to recover the fraction used in the TAF-DPS. Unlike the case of PLL, this approach is in direct fashion that can result in lower-cost-implementation. Table 5.7 is the frequency plan. In this plan, $N = 8$. This N divider can generate 16 outputs ($K = 16$) of evenly distributed phases from the in-phase (0°) and out-phase (180°) source signals. Using the configuration of

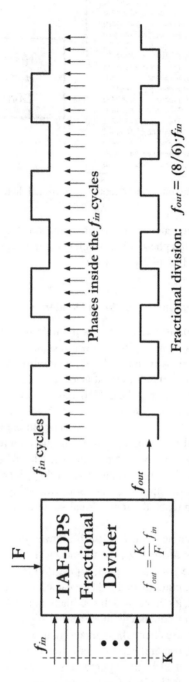

FIGURE 5.45 TAF-DPS can function as a fractional divider.

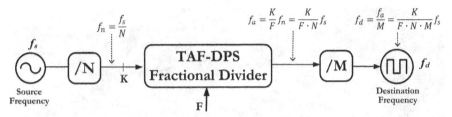

FIGURE 5.46   Scheme using TAF-DPS to achieve fractional divide ratio in direct fashion.

Figure 5.46, the frequency at the TAF-DPS output is always at 256 MHz ($f_a = 256$ MHz). This signal, however, is in time-average-frequency fashion since a fraction is used in F (refer to column 4 in Table 5.7). But this plan is designed in such way that all the fractions are 1/128 based. Further, the post divider M always takes the value of 128. It serves two purposes: (a) to recover the 1/128-based fractions and (b) to generate the final output at exactly 2 MHz. For this plan, the ratio of f_s/f_a can be derived as $f_s/f_a = (F{\cdot}N/K) = F/2$ and the values are listed in column 6 of Table 5.7. The ratio of f_s/f_d is $f_s/f_d = F{\cdot}64$. For all the 16 channels in Table 5.7, the corresponding F value in 4th column will make $f_d = 2$ MHz.

A 2.4 GHz ISM band wireless senor system is designed in a 65 nm CMOS process for a 250 Kb/s date rate. The source signal generator is an LC DCO which is used to generate the RF signal for the 16 channels. This DCO is always in the free-run mode. Besides the 16 RF frequencies, it is also used to generate the baseband 2 MHz signal using the scheme proposed in Figure 5.46. The picture on the left in Figure 5.47 shows the laboratory configuration to test this open-loop divider chain. The outputs from the DCO and the TAF-DPS are fed to channels 1 and 3, respectively, of an

TABLE 5.7   Frequency Plan Using TAF-DPS to Achieve Fractional Division

Channel Number	f_s (MHz)	f_n (MHz)	Control Word $F = I + r$	f_a (MHz)	$f_s/f_a = F/2$	f_s/f_a Measured
11	2405	300.625	$18 + 101/128$	256	9.39453125	9.3925781
12	2410	301.25	$18 + 106/128$	256	9.41406250	9.4121094
13	2415	301.875	$18 + 111/128$	256	9.43359375	9.4316407
14	2420	302.5	$18 + 116/128$	256	9.45312500	9.4511719
15	2425	303.125	$18 + 121/128$	256	9.47265625	9.4707030
16	2430	303.75	$18 + 126/128$	256	9.49218750	9.4902344
17	2435	304.375	$19 + 3/128$	256	9.51171875	9.5097656
18	2440	305	$19 + 8/128$	256	9.53125000	9.5292968
19	2445	305.625	$19 + 13/128$	256	9.55078125	9.5488281
20	2450	306.25	$19 + 18/128$	256	9.57031250	9.5683594
21	2455	306.875	$19 + 23/128$	256	9.58984375	9.5878905
22	2460	307.5	$19 + 28/128$	256	9.60937500	9.6074219
23	2465	308.125	$19 + 33/128$	256	9.62890625	9.6269531
24	2470	308.75	$19 + 38/128$	256	9.64843750	9.6464844
25	2475	309.375	$19 + 43/128$	256	9.66796875	9.6660156
26	2480	310	$19 + 48/128$	256	9.68750000	9.6855469

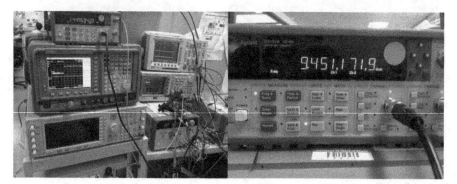

FIGURE 5.47   Test configuration (left) and measured divide ratio for channel 14 (right).

Agilent 53131A frequency counter. The ratio f_s/f_a is consequently measured. The picture on the right shows the f_s/f_a test result for channel 14. The test results for all the channels are included in column 7 of Table 5.7. Those numbers show a high level of agreement between the calculation and the measurement with a systematic error of ~0.001953 for all control settings. Figure 5.48 plots the calculated and the measured results. A high degree of correlation is clearly visible. The systematic error could be due to the mismatch between channels 1 and 3 of this frequency counter.

5.12.3   TAF-DPS in Pulse Width Modulation

Pulse width modulation (PWM) is an important circuit technique that has many applications, such as power delivery, voltage regulation, class-D audio amplifier, and pulse code modulation (PCM) digital sound. A PWM pulse train can be generated by a microcontroller unit (MCU) controlled counter or by an RC delay based analog PWM modulator. The TAF-DPS is a circuit technique that can also function as a PWM generator since it directly constructs each pulse in its output pulse train. As illustrated in Figure 5.49, it can produce three types of pulse trains: type I of fixed duty cycle with varying period, type II of varying duty cycle with fixed period, and type III of fixed pulse length with varying period. Those pulse trains can be conveniently generated from one TAF-DPS PWM generator through a PWM control

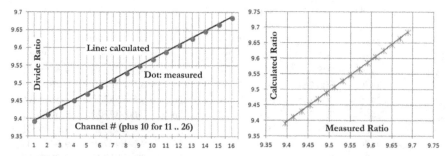

FIGURE 5.48   Calculated and measured results: divide ratio vs. channel number (left); calculated ratio vs. measured ratio (right).

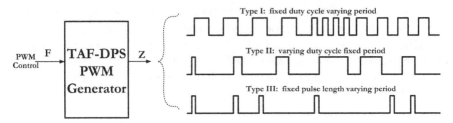

FIGURE 5.49 TAF-DPS as PWM generator for generating three types of pulse train signals.

input. Figure 5.50 shows the transistor-level simulation result from a TAF-DPS circuit designed in a 0.18 μm CMOS process. From top to bottom, the three types of pulse trains are shown. In each case, the bottom trace is the PWM waveform, and the top trace is the control command. As seen, the PWM waveforms follow the patterns in the control commands.

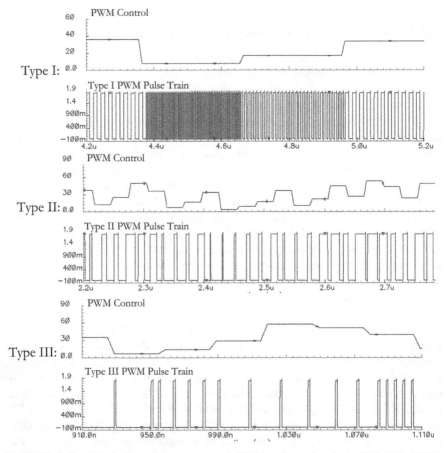

FIGURE 5.50 Simulation of three types of PWM pulse trains from TAF-DPS PWM generator.

Compared to the MCU-based counter PWM and the *RC* delay analog PWM, the TAF-DPS PWM technique is highly flexible. Since a PWM circuit is mainly used in low-MHz applications, the TAF-DPS PWM generator can be constructed purely on digital standard cells. This can lead to very low-cost and low-power implementation. It can be a very useful tool for implementing designs in many emerging applications, such as the Internet of Things (IoT).

5.12.4   TAF-DPS for Message Transmission Using Spectrum

From the discussion in previous sections and in [Xiu12a], it can be seen that the TAF-DPS output is very rich in its spectrum content. Further, the spectrum and the frequency control word have a fixed and predictable relationship. This feature can potentially be used for message transmission, a illustrated in Figure 5.51. This is possible because (a) there is a one-to-one deterministic relationship between the control word and the spectrum and (b) the spectrum is precisely calculable (refer to Chapter 5 of [Xiu12a]).

5.12.5   TAF-DPS for Message Transmission Using Duty Cycle

Data transmission is a critical task in modern chip design. As discussed in previous sections, CDR is an elegant solution that combines the data and clock into one communication channel. It enables high-speed data communication in serial fashion, which eliminates the clock–data skew problem. It also lowers the overall system cost by reducing the number of channels. It however requires sophisticated recovery circuitry on the receiver side. It also demands that the transmitter encode enough transitions in the transmitting data stream. A more straightforward solution of combining the clock and data is to create an electrical signal such that the transition edge of the pulse represents the clock and its duty cycle represents the data. Figure 5.52 illustrates this idea. On the left is the conventional NRZ (non-return-to-zero) digital data signal. The top trace is the clock and the bottom is the data. On the right is the ECDD (edge for clock and duty for data) signal where the rising (or falling) edge is the clock signal and the duty cycle represents the data. In each ECDD pulse, the waveform is made in such a way that its duty cycle is significantly different from 50%. The low and high portions are either $x\%$ and $(1 - x)\%$ or $(1 - x)\%$ and $x\%$, depending on the value of the digital data.

TAF-DPS is an ideal tool for creating the ECDD signal since each pulse can be precisely constructed using the base time unit Δ. Using TAF-DPS, both the pulse length and the duty cycle can be synthesized by fixed values of Δ. The key design consideration in this architecture is the determination of the x value (i.e., the low–high ratio). The smaller the x is, the larger the low–high difference is (and thus the easier it is to design a circuit to differentiate them). However, smaller x values lead to narrower pulse, which results in higher risk of being swallowed during transmission. Table 5.8 lists some implementation options for a ECDD signal of 400 MHz. A TAF-DPS based on a four-differential-stage ring oscillator (eight outputs) can support all the solutions in this table.

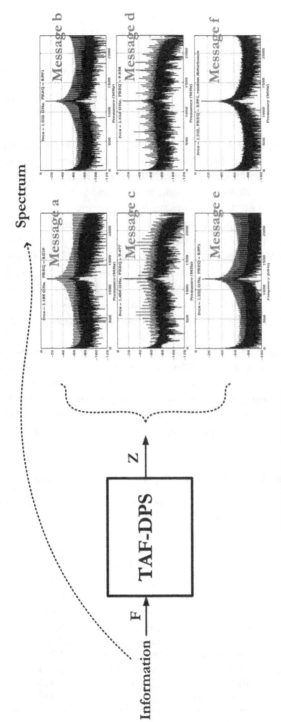

FIGURE 5.51 TAF-DPS for message transmission using spectra.

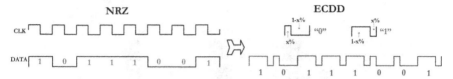

FIGURE 5.52   Conventional NRZ digital signal (left) and ECDD signal (right).

On the receiving side, the ECDD signal can be decoded by a circuit made of multiple samplers. Each sampler can be driven by a clock from a plurality of clock signals which are evenly spaced in one ECDD cycle. The resultant bit stream can be fed into a voting circuit. Based on the number of 0s and 1s received, the circuit is able to recognize whether a digital 0 or 1 is sent from the transmitter. This architecture of an ECDD one-wire data communication system is illustrated in Figure 5.53 [Xiu15]. In the TX, the ECDD modulation circuit is a TAF-DPS. A PLL or DLL is used to generate the multiple outputs required by the TAF-DPS. The data latch circuit captures the incoming NRZ data by using the ECDD signal (or a reconditioned signal of same frequency) as the clock. The resultant DATA_T is fed to TAF-DPS and is used to control the duty cycle. On the RX side, the ECDD signal is first used as a reference for a PLL (or DLL) which generates the multiple clock signals for the samplers. At the same time, the ECDD signal is fed to a demodulation circuit, which could be a majority voting circuit as described above. The resulting DATA_R is the received data, which is accompanied by its clock CLK_R (one of the outputs from the PLL/DLL).

Figure 5.54 shows the simulation result of an ECDD one-wire communication system using a circuit constructed according to Figure 5.53 in a 0.18 μm CMOS process. The ECDD signal is designed at a frequency of 300 MHz. A four-differential-stage VCO is used to generate eight outputs at 300 MHz for the TAF-DPS ($\Delta = 0.417$ ns).

TABLE 5.8   Various Plans for Constructing 400 MHz ECDD Signal

Pulse Length: 2.5 ns	Low Portion	Low–High Difference	Minimal Pulse Length (ns)	x (%)
8Δ ($\Delta = 0.3125$ ns)	5Δ (or 3Δ)	$2\Delta = 0.625$ ns	0.9375	37.5
8Δ ($\Delta = 0.3125$ ns)	6Δ (or 2Δ)	$4\Delta = 1.25$ ns	0.625	25
10Δ ($\Delta = 0.25$ ns)	6Δ (or 4Δ)	$2\Delta = 0.5$ ns	1	40
10Δ ($\Delta = 0.25$ ns)	7Δ (or 3Δ)	$4\Delta = 1$ ns	0.75	30
10Δ ($\Delta = 0.25$ ns)	8Δ (or 2Δ)	$6\Delta = 1.5$ ns	0.5	20
12Δ ($\Delta = 0.2083$ ns)	7Δ (or 5Δ)	$2\Delta = 0.417$ ns	1.042	42
12Δ ($\Delta = 0.2083$ ns)	8Δ (or 4Δ)	$4\Delta = 0.833$ ns	0.833	33
12Δ ($\Delta = 0.2083$ ns)	9Δ (or 3Δ)	$6\Delta = 1.25$ ns	0.625	25
14Δ ($\Delta = 0.1786$ ns)	8Δ (or 6Δ)	$2\Delta = 0.357$ ns	1.07	43
14Δ ($\Delta = 0.1786$ ns)	9Δ (or 5Δ)	$4\Delta = 0.714$ ns	0.893	36
14Δ ($\Delta = 0.1786$ ns)	10Δ (or 4Δ)	$6\Delta = 1.07$ ns	0.714	29
14Δ ($\Delta = 0.1786$ ns)	11Δ (or 3Δ)	$8\Delta = 1.43$ ns	0.536	21

Note: Bit time = 2.5 ns.

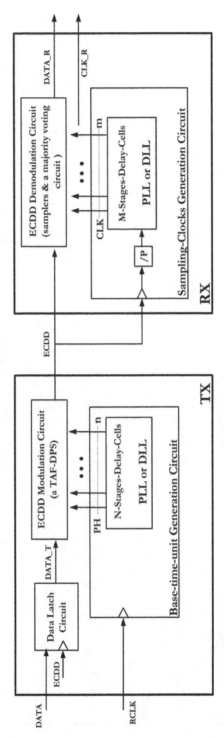

FIGURE 5.53 Architecture of ECDD one-wire data communication system.

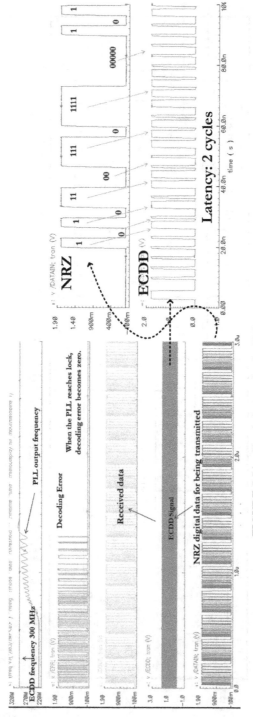

FIGURE 5.54 Simulation result of ECDD one-wire data communication.

The VCO is locked to a reference through a PLL. The pulse length of the ECDD signal is 8Δ. The duty cycle value x is chosen as 25% (i.e., the ratio is 6Δ to 2Δ). On the left side, the bottom trace is a sequence of digital data in NRZ digital format. The second trace from the bottom is its corresponding ECDD signal generated by the ECDD modulation circuit. The third trace from the bottom is the output data from the ECDD demodulation circuit (the DATA_R in Figure 5.53). The fourth trace from the bottom is the output from an error-checking circuit. When the received data have error, the error-checking circuit output is in logic level high. Otherwise, it is in logic level low. The plot at the very top has two traces. The trace of the PLL output frequency is the frequency measurement of one of the sampling clocks generated by the PLL in the receiver (inside the RX). The trace of the frequency embedded in the ECDD signal is the frequency measurement of the ECDD signal.

In this simulation, the original digital data are generated by a nine-element PRBS (pseudorandom binary sequence) encoder using a driving clock of 300 MHz. The error-checking circuit is made of its corresponding nine-element PRBS decoder. Thus, the inherent logics (randomness) of both the PRBS encoder and decoder match. As a result, the error-checking circuit will report no error if the data from the transmitter are received correctly by the receiver. In this simulation, the error-checking circuit is clocked by one of the PLL outputs (inside the RX). As shown, when the PLL reaches lock, the sampling clock frequency matches that of ECDD signal (300 MHz, the same as the PRBS encoder frequency). Furthermore, after the PLL reaches lock, the decoder output (the error signal) becomes logic level low. This indicates that the sampling circuit and the majority voting circuit correctly receive and decode the ECDD signal. On the right-hand side, the zoom-in versions of the original data waveform and its corresponding ECDD waveform are shown. The NRZ signal is at the top and the corresponding ECDD signal is at the bottom with a delay of two cycles. This delay is consumed by the circuit for processing (the latency).

REFERENCES

[Ben89]   A. Benjaminson and S. Stallings, "A microcomputer-compensated crystal oscillator using dual-mode resonator," *Proc. 43rd Annu. Symp. Freq. Control*, pp. 20–26, 1989, IEEE Catalog No. 89CH2690-6.

[Ben91]   A. Benjaminson and B. Rose, "Performance tests on an MCXO combining ASIC and hybrid construction," *Proc. 45th Annu. Symp. Freq. Control*, pp. 393–397, 1991, IEEE Publication No. 91CH2965-2.

[Bje06]   T. Bjerregaard and S. Mahadevan, "A survey of research and practices of network-on-chip," *ACM Comput. Surv.*, vol. 38, Article 1, March 2006.

[Blo89]   M. Bloch, M. Meirs, and J. Ho, "The microcomputer compensated crystal oscillator (MCXO)," *Proc. 43rd Annu. Symp. Freq. Control*, pp. 16–19, 1989, IEEE Catalog No. 89CH2690-6.

[Che04]   T. Chelcea and S. M. Nowick, "Robust interfaces for mixed-timing systems," *IEEE Trans VLSI Syst.*, vol. 12, no. 8, pp. 857–873, Aug. 2004.

[Fer13]   J. L. Ferrant et al., *"Synchronous Ethernet and IEEE 1588 in Telecomes,"* Wiley, Hoboken, NJ, June 2013.

[Fil89]   R. Filler and J. Vig, "Resonators for the microcomputer-compensated crystal oscillator," *Proc. 43rd Annu. Symp. Freq. Control,* pp. 8–15, 1989, IEEE Catalog No. 89CH2690-6.

[Gri14]   D. Griffith, "Low power timer for wireless networks," IEEE CAS presentation, University of Texas, Dallas, TX, May 2014.

[Hor04]   M. Horauer, "Clock synchronization in distributed systems," Ph.D. Thesis, Vienna University of Technology, 2004.

[Int04]   Intel, "Enhanced Intel SpeedStep technology for the Intel Pentium M Processor," Whiter Paper 301170-001, Intel Corp., 2004. http://download.intel.com/design/network/papers/30117401.pdf

[Jac96]   E. Jackson, H. Phillips, and B. E. Rose, "The micro-computer compensated crystal oscillator—A progress report," *Proc. 1996 IEEE Int. Freq. Control Symp.,* pp. 687–692, 1996, IEEE Publication No. 96CH35935.

[Ken08]   D. Kenny and R. Henry, "Comparative analysis of MEMS, programmable and synthesized frequency control devices versus traditional quartz based devices," *Proc. IEEE Int. Freq. Control Symp.,* pp. 396–401, May 2008.

[Kol09]   A. Kolodny, "Networks on chips: A new paradigm," Tutorial, 2009.

[Kuh96]   T. S. Kuhn, *The Structure of Scientific Revolutions,* 3rd ed., University of Chicago Press, Chicago, IL, Dec. 1996.

[Mas05]   N. Masakatsu et al., "Dynamic voltage and frequency management for a low-power embedded microprocessor," *IEEE J. Solid-State Circuits,* vol. 40, pp. 28–35, Jan. 2005.

[Mcc09]   M. S. McCorquodale et al., "A 25-MHz self-referenced solid-state frequency source suitable for XO-replacement," IEEE J. Solid-State Circuits, pp. 943–956, May 2009.

[Mes90]   D. G. Messerschmitt, "Synchronization in digital system design," *IEEE J. Selected Areas Commun.,* no. 8, pp. 1404–1409, Oct. 1990.

[Mul05]   R. Mullins, "Asynchronous vs. synchronous design techniques for NoCs," tutorial, International Symposium on System-on-Chip, Tampere, Finland, Nov. 2005.

[Nat09]   National Instruments, "FlexRay automotive communication bus overview," National Instruments, Aug. 2009.

[Ngu99]   C. T.-C. Nguyen and R. T. Howe, "An integrated CMOS micromechanical resonator high-Q oscillator," *IEEE J. Solid-State Circuits,* vol. 34, no. 4, pp. 440–455, Apr. 1999.

[Ran98]   N. Ranganathan, N. Vijaykrishnan, and N. Bhavanishankar, "A linear array processor with dynamic frequency clocking for image processing applications," *IEEE Trans. Circuits Syst. Video Technol.,* vol. 8, pp. 435–445, Aug. 1998.

[Sch03]   U. Schmid, "Interval-based clock synchronization with optimal precision," *Information and Computation,* vol. 186, no. 1, pp. 36–77, 2003.

[Sch89]   S. Schodowski, "Resonator self-temperature-sensing using a dual-harmonic-mode crystal oscillator," *Proc. 43rd Ann. Symp. Freq. Control,* pp. 2–7, 1989, IEEE Catalog No. 89CH2690-6.

[Sin12]   N. Sinoussi et al., "A single LC tank self-compensated CMOS oscillator with frequency stability of +/– 100 ppm from −40°C to 85°C, *Proc. 2012 IEEE Int. Freq. Control Symp.,* pp. 1–5, 2012.

[Sun06]   K. Sundaresan, P. E. Allen, and F. Ayazi, "Process and temperature compensation in a 7-MHz CMOS clock oscillator," *IEEE J. Solid-State Circuits*, pp. 433–442, Feb. 2006.

[Tee07]   P. Teehan et al., "A survey and taxonomy of GALS design styles," *IEEE Des. Test Comput.*, vol. 24, no. 5, pp. 418–428, Sept.–Oct. 2007.

[Val10]   V. Stojanović, "Design of energy-efficient on-chip networks," ISSCC 2010 Tutorial.

[Vig01]   J. R. Vig, "Quartz crystal resonators and oscillators for frequency control and timing applications—A tutorial," Rev. 8.5.1.2, AD-M001251, March 2004. file:///C:/Users/Liming/Downloads/quartz_crystal_resonators_and_oscillators___ for_frequency_control_and_timing_applications.pdf

[Vog13]   K. De Vogeleer, G. Memmi, P. Jouvelot, and F. Coelho, "The energy/frequency convexity rule: Modeling and experimental validation on mobile devices," Technical Report 2013D008, TELECOM ParisTech, Sept. 2013. http://www.mathpubs.com/detail/1401.4655v1/The-EnergyFrequency-Convexity-Rule-Modeling-and-Experimental-Validation-on-Mobile-Devices

[Xiu07]   L. Xiu, S. Clynes, S. Gurrapu, T. Haider, F. Ying, and W. Mohammed, "Flying-adder PLL based synchronization mechanism for data packet transport," Paper presented at 2007 IEEE Dallas/CAS Workshop, 2007. System-on-Chip, 2007. DCAS 2007. 6th IEEE Dallas Circuits and Systems Workshop on.

[Xiu08]   L. Xiu, "A novel DCXO module for clock synchronization in MPEG2 transport system," *IEEE Trans. Circuit Sys. I*, vol. 55, pp. 2226–2237, Sept. 2008.

[Xiu12a]   L. Xiu, *Nanometer Frequency Synthesis beyond Phase Locked Loop*, Wiley-IEEE Press, Hoboken, NJ, Aug. 2012.

[Xiu12b]   L. Xiu, K.-H. Lin, and M. Ling, "The impact of input-mismatch on flying-adder direct period synthesizer," *IEEE Trans. Circuit Syst. I*, vol. 59, pp. 1942–1951, Sept. 2012.

[Xiu14a]   L. Xiu, "Direct period synthesis for achieving sub-PPM frequency resolution through time average frequency: The principle, the experimental demonstration, and its application in digital communication," *IEEE Trans. VLSI*, accepted for publication, June 2014; available at IEEE Xplore.

[Xiu14b]   L. Xiu, "Circuit and method of using time-average-frequency direct period synthesizer for improving crystal-less frequency generator's frequency stability," U.S. patent 8890591, Nov. 2014.

[Xiu15]   L. Xiu, "Circuits and methods for one-wire communication bus of using pulse-edge for clock and pulse-duty-cycle for data," U.S. patent 8929467, Jan. 2015.

[Yan12]   F. Yang and T. Haider, "MPEG-2 transport stream packet synchronizer," U.S. patent 8249171, 2012.

6

TAF-DPS CLOCK GENERATOR AND ON-CHIP CLOCK DISTRIBUTION

6.1   GLOBAL CLOCK DISTRIBUTION AT LOW FREQUENCY

Clock distribution is a critical task in modern chip design. In recent years, advances in CMOS technology have led to an exponential increase in chip complexity. The number of transistors in large chips has reached billions [Tam09]. Modern SoC can be regarded as many on-chip micronetworks communicating to each other all the time. The clock signal is the key that makes this happen. From a clocking perspective, chip architecture can be classified as GALS and globally synchronous locally synchronous (GSLS). In the GSLS approach, unlike GALS, the clock signals driving all the on-chip modules run at the same frequency. Among them, they also have a fixed phase relationship. This requires the distribution of a global clock signal. There are several design considerations when distributing a clock signal globally: the skew caused by different distribution paths, the jitter accumulated along the distribution path, the silicon and metal resource required for routing the clock signal, and the power used by the distribution network. Figure 6.1 illustrates the clock distribution methods commonly used in chip design.

In Figure 6.1a, conventional tree structures are used to distribute the clock signal. The distribution network can be constructed using a branch tree, H-tree, or X-tree. In the illustration of Figure 6.1b, an H-tree with active skew compensation is depicted. To alleviate the skew problem, the delays at the ends of different branches are compared. The result is used to drive delay lines so that the delays of the paths can be adjusted. Consequently, skew can be minimized. Clock mesh (clock grid) is also used for some

From Frequency to Time-Average-Frequency: A Paradigm Shift in the Design of Electronic Systems, First Edition. Liming Xiu.
© 2015 The Institute of Electrical and Electronics Engineers, Inc. Published 2015 by John Wiley & Sons, Inc.

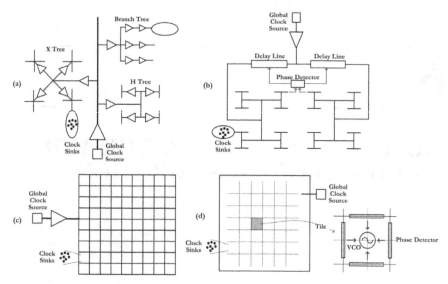

FIGURE 6.1 Clock distribution methods: (a) conventional trees, (b) delay/skew-compensated H-tree, (c) clock mesh, and (d) distributed PLL array.

designs, especially in high-end microprocessor. In this method, a solid grid made of metals is constructed on-chip as shown in Figure 6.1c. Its purpose is to deliver the clock signal to all the locations in the chip. In practice, the tree and grid methods can be used together to achieve the goal of delivering a clock signal from a source to all the sinks across a large chip [Cha09]. To minimize skew actively, using a distributed PLL array is proposed [Pra95, Gut00, Zia13]. In Figure 6.1d, the entire chip is split into multiple small areas called tiles. Inside each tile, there is a local frequency generator (represented by the VCO symbol). Along the four boundaries of each tile phase detectors are used to compare the delay differences between the local clock and its neighboring clocks. The result is used to drive the frequency generator and then to minimize the skew. In this approach, the array of distributed PLLs actively compensates the skew.

As semiconductor process technology advances, the tree structure faces difficult challenges. The circuit operating frequency becomes higher due to the reduction in transistor gate delay. The chip size becomes larger since more transistors can be packed. As a result, the global clock signal has to travel further. Moreover, both the gate and interconnect delay variations induced by PVT (Process, Voltage, Temperature) change become larger. Furthermore, the interconnect delay does not scale well with process advance. All these factors have made skew take a larger percentage of the clock period. They also make it hard to control the variation in skew. To make it even worse, delivery of the clock signal across the chip at high frequency requires large amounts of metal resource (for shielding) and high consumption of energy (could be as high as 50% of the total power used by the chip). The distributed PLL array also has the high-resource and high-power-consumption problems. Further, it has a stability problem due to the fact that many PLLs must lock to the same common reference.

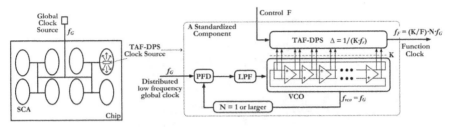

FIGURE 6.2   TAF-DPS-enabled clock distribution of low global frequency.

The TAF-DPS clock source explained in Section 3.4 provides a new possibility for solving the clock distribution problem. As shown in Figure 3.5 and expressed in $f_{out} = (K/F)f_r$, the TAF-DPS is able to generate frequency that is higher than its input. Furthermore, as explained in Section 4.25 of [Xiu12], TAF-DPS can control its output phase. In other words, the phase of its output signal has a fixed relationship with the input reference signal and the phase is adjustable if desired. This enables a distribution approach using a low frequency in the global clock. When reaching destinations, the low-frequency global clock can be frequency boosted by a local TAF-DPS to generate the function clock. This scheme is depicted in Figure 6.2.

As shown on the left in Figure 6.2, the chip is split into multiple synchronous clocking areas (SCAs). Each SCA is small enough that a conventional clock distribution method (such as a tree) can be efficiently used to deliver the clock within the area. The goal of those SCAs is synchronous (i.e., running at the same frequency with fixed phase relationship to each other). Within each SCA, one TAF-DPS clock generator is used to generate the function clock to support local operation. Its structure is depicted on the right-hand side of Figure 6.2. A 1x PLL is used to generate the K outputs to support the TAF-DPS. The distributed low-frequency global clock signal f_G is used as its reference. The function clock f_F is generated at frequency $f_F = (K/F)f_G$, where F is the frequency control word. As seen, when $F < K$, we can generate a function clock with higher frequency.

Figure 6.3 includes two transistor-level simulation results using the same f_G. In the left simulation, the global clock is delivered to their respective PLLs for both SCA#1 and SCA#2. In both PLLs, one of the K VCO outputs is phase aligned with the global clock. Therefore, the two VCO outputs are aligned to each other even though they are in different SCAs. Within each TAF-DPS, the circuit guarantees that its output has a fixed and known phase relationship with that particular VCO output. This is demonstrated in the simulations of Figure 6.3. As a result, the outputs of both TAF-DPS (the two function clocks associated with those two SCAs) have same frequency and are phase aligned. This ensures the synchronous operation of the system (GSLS). The simulation on the right-hand side shows the case of generating another functional frequency. In both cases, the function clocks have higher frequency than that of the global clock (much higher frequency can be generated but it would be difficult to display clearly).

Figure 6.4 shows the feature that allows the TAF-DPS to adjust its output phase. If SCA#1 and SCA#2 are required to have different phases (such as for imbalanced

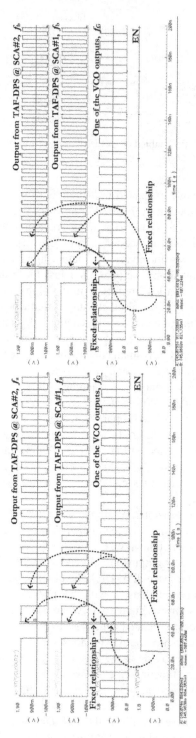

FIGURE 6.3 TAF-DPS clock sources in all SCAs are synchronous.

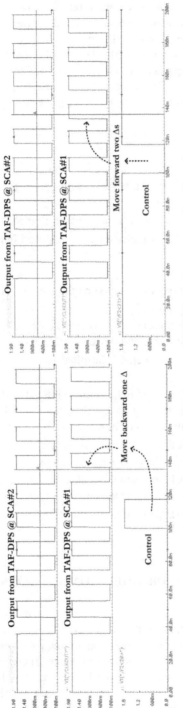

FIGURE 6.4 Phase relationship between TAF-DPS clock sources can be adjusted.

delays in their global clock delivery paths), one TAF-DPS can adjust its phase to fulfill the requirement. This is because, as explained in Section 4.25 of [Xiu12], the TAF-DPS can be viewed as a DLL with infinite delay adjustment capability. In the simulation on the left the output from the TAF-DPS of SCA#1 is moved backward by 1 Δ. In the one on the right, it is moved 2 Δ forward. This application scenario can happen when there is data communication between SCA#1 and SCA#2. This phase adjustment capability can help align the data and clock appropriately so that data can be captured more reliably.

The features of the TAF-DPS-enabled low global frequency distribution method are summarized as follows:

- All SCA function clocks can have the same frequency (synchronous) or different frequencies if desired.
- All SCA function clocks can have their phases aligned. This leads to zero skew.
- If desirable, each SCA clock phase can be adjusted individually (useful skew between SCAs).

The advantages of this clock distribution method are as follows:

- Low frequency can be used in the global clock signal. This leads to significant reduction in routing resource and power consumption.
- The skew problem can be better controlled since the global clock is in low frequency.
- The function clock is expected to have lower jitter since less jitter accumulation is experienced in the delivery path. Also, the 1x PLL can filter out all the high-frequency jitter collected from the delivery process.
- This method scales well with the process technology since the performance of TAF-DPS improves with the advances in process.
- The PLL in each SCA is individually locked to the global clock. Thus, there is no stability problem (there is no inter-module clock interaction).
- Less electromagnetic (EM) radiation is generated since the global clock is in low frequency.

6.2   RESONANT CLOCK DISTRIBUTION NETWORK ASSISTED BY TAF-DPS

The most characteristic feature of the clock network is the large capacitive loading that it presents to the clock source. During operation, the clock source is responsible for the charge and discharge of this large capacitance. Therefore, an effective way of lowering power is to recycle the energy used by charging and discharging this large capacitance in the clock network (charge recovery clock distribution). This approach is realized by using the principle of *LC* resonance in clock distribution, as illustrated

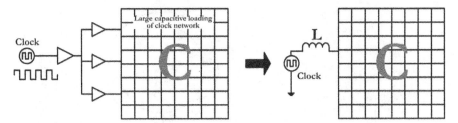

FIGURE 6.5   Principle of energy recycle in *LC* resonance clock distribution.

in Figure 6.5. Inductance is intentionally introduced into the chip structure. The large capacitance associated with the clock distribution network functions as the *C* of the *LC* oscillator. During the charge and discharge process, the energy is stored and released periodically. Ideally, 100% of the energy can be recycled and the electrical oscillation (the clock waveform) can be self-sustained. In practice, due to the parasitic resistance associated with the inductor and the clock sinks, some portion of the energy is lost as the generated heat. Hence, compensation circuitry has to be incorporated on-chip to provide the energy supporting the oscillation. It is expected that the energy used by this approach is much lower than that consumed in conventional methods since, instead of CV^2f, the consumed power now is I^2R where R is the total parasitic resistance. This power is frequency independent and hence this approach is a good candidate for distributing the clock in high frequency (GHz range).

Figure 6.6 illustrates four methods reported in the literature of delivering a clock signal in an *LC* resonance circuits. In Figure 6.6a, the clock distribution network is divided into multiple clock sectors [Cha03, Cha04, Cha09]. The global clock signal is delivered to each sector using an H-tree. Within each sector, the clock signal is further delivered to multiple (e.g., four) locations. In each such location, there is a spiral inductor whose two ends are connected to the global clock network and a decoupling capacitor, respectively. The decoupling capacitors, the parasitic capacitance associated with the clock network, and the inductors form an *LC* resonant circuit that is used to recycle some of the clock energy. In this structure, however, the actual clock sinks (flip-flops, latches, etc.) are driven by local clock buffers attached to a clock grid. This grid is connected to and driven by the global *LC* distribution network (see Figure 6.6a). Therefore, the large capacitance associated with the clock loadings is isolated from the *LC* network and does not participate in the *LC* resonance. This leads to reduced efficiency in recycling clock power.

In Figure 6.6b, the clock *LC* distribution network directly reaches the clock sinks (no clock buffer in the network). Thus, all the parasitic capacitance takes part in the *LC* resonance. It results in a larger degree of power saving [Sat08]. However, since the *LC* network outputs a sinusoidal waveform, the slow rising and falling clock edges limit the clocking speed. Moreover, due to the lack of isolation between the *LC* network and the clock elements, the data-induced jitter will significantly degrade the quality of the clock signal. Furthermore, the popular clock gating technique is hard to implement due to the elimination of the buffers in the delivery chain.

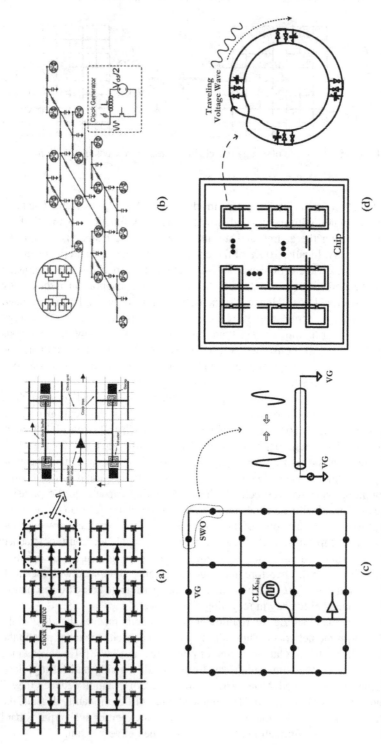

FIGURE 6.6 Clock distribution using *LC* resonance: (a) *LC* resonance at clock sector [Cha04, Cha09], (b) *LC* resonance at actual clock loading [Sat08], (c) standing-wave *LC* clock distribution network [Fra03b], (d) traveling wave clock generation and distribution network [Woo01].

When two waves of the same frequency travel in opposite directions and interact with each other, standing and traveling waves will be formed. Equation (6.1) demonstrates the result when two waves $A\cos(\omega t - \beta z)$ and $B\cos(\omega t + \beta z + \phi))$ travel in opposite directions within the same transmission line:

$$A\cos(\omega t - \beta z) + B\cos(\omega t + \beta z + \phi)$$
$$= 2B\cos(\omega t + \tfrac{1}{2})\cos(\beta z + \tfrac{1}{2}) + (A - B)\cos(\omega t - \beta z) \tag{6.1}$$

where t and z are independent variables for time and location, respectively. The resultant first term is the standing wave, which has uniform phase in all locations along the line. Its amplitude is location dependent. The second term is the traveling wave, whose phase is location dependent but its amplitude is constant.

Since a standing wave has the unique feature of uniform phase, it can be used for clock distribution with potentially zero skew. In Figure 6.6c, a clock distribution network is constructed by a metal structure with multiple virtual grounds (VGs). The VGs are formed by shorting them together. Between two VGs, a standing wave oscillator (SWO) is formed since the electrical waves are reflected at the VGs. This leads to the interaction of waves, resulting in a standing wave [the traveling wave is minimized by the fact that the two wave amplitudes are almost equal, or $A \approx B$ in (6.1)]. A clock signal is injected into the network to initialize and maintain the oscillation. Clock buffers are attached to the metal structure to form the local clock distribution network [Fra03]. In another design, the VGs are attached by inductors to alleviate the amplitude variation along the line (the amplitudes no longer go to zero at the ends) [Mam09].

Figure 6.6d shows how a traveling wave is used to generate a clock signal and distribute the clock signal across a chip. The traveling wave is produced by connecting one end of the transmission line to the other so that a loop is formed [Woo01]. As shown, instead of tying the ends of the transmission line to virtual ground, the ends are cross connected (Möbius ring). In this way, no reflection results and the standing wave is nonexistent [$B = 0$ in (9.1)]. Multiple antparallel inverter pairs are added to the line to compensate for the energy lost and maintain rotation lock. The traveling wave has a constant amplitude but different phases along the points on the path. After startup, the wave can travel in either direction (usually the direction of lowest lost) unless it is intentionally biased (directional coupling) [Zha09]. The oscillation frequency can be slightly adjusted by adding tunable capacitors between the line. This traveling wave can be distributed to the whole chip as the global clock signal by connecting multiple such structures together, as illustrated in the Figure 6.6d.

The LC resonance clock distribution methods described above have great potential to lower the power consumption of the clock network. They all, however, lack the frequency flexibility required in microprocessor operation. This is because the oscillation frequency (for standing waves and traveling waves) or optimal working frequency (for the structures of Figures 6.6a and 6.6b to achieve maximum power saving) is determined by the physical structure of the network (i.e., the natural frequency of the LC resonator). As a circuit technique, in supporting LC resonance clock generation and distribution, TAF-DPS can be used to enhance frequency flexibility. When the time-average-frequency concept is adopted, the frequency generation capability can be further expanded into the sub-ppm range.

Of the four *LC* resonance methods discussed above, using the standing wave oscillator (Figure 6.6c) or traveling wave oscillator (Figure 6.6d) is the most promising ones. The reason is that the clock quality (skew and accumulated jitter) can be better controlled since the global clock is distributed by chip-size distributed *LC* resonators. In both approaches, the clock generation (frequency synthesis) task can be accomplished by TAF-DPS. In Figure 6.7, a scheme of using SWO and TAF-DPS for clock distribution and clock generation is illustrated. As described previously, all the locations in the SWO ring can provide clock signals of the same frequency and same phase. Usually, this SWO ring can operate in high frequency, such as ~10 GHz in a 0.13 µm process. Further, the output is available as a differential pair. This high oscillating frequency provides an opportunity for generating multiple outputs through dividers. For example, a divider chain made of 8 CML (Current Mode Logic) dividers can produce 16 evenly distributed phases from this *LC* resonator output pair, as shown in the figure. This group of phases can be used to support the TAF-DPS operation. Each region shown in the figure can have a dedicated TAF-DPS. The task of frequency synthesis on a functional clock is accomplished by those TAF-DPSs. As discussed in Section 6.1, the TAF-DPS outputs can all be made synchronous to each other since their inputs are synchronized by the SWOs, resulting in zero skew (theoretically).

The traveling wave oscillator (RTWO) clock distribution depicted in Figure 6.6d is further illustrated in Figure 6.8. The *LC* transmission line is configured as a Möbius ring. Within each ring, signal phases are uniformly distributed, as shown on the left. When tapered out, these phases naturally provide the input for TAF-DPS operation. There is, however, a synchronization problem when these rings are connected together as an ROA (rotary oscillator array). For a single ring, the wave traveling direction is the one with less impedance. It could be clockwise or counterclockwise. When connected together as shown in the middle in Figure 6.8, the neighboring rings could oscillate at different directions (nondeterministic). This leads to difficulty in synchronizing the clocks around the chip because the same phase point of all the rings has to be tested beforehand.

In [Ten11], an elegant solution is proposed by arranging RTWO rings in a different topology called an ROA brick. The basic ROA brick structure is illustrated on the left in Figure 6.9. Compared to the structure in the middle of Figure 6.8, this one is its "mirror structure" topologywise. The important feature of this structure is that the traveling wave in all the associated rings travels in the same direction, either all clockwise or all counterclockwise. This fact makes it possible to identify the same phase point on all the rings, which is labeled in the drawing. Chipwise, the ROA bricks can be arranged as shown on the right. Every four RTWO rings make a ROA brick and every ROA brick shares two RTWO rings with its neighbor. In this way, the same phase point can be established around the whole chip.

Starting from the same phase point, other locations can be tapered out to construct the multiple phases needed for TAF-DPS operation. The sequential order is

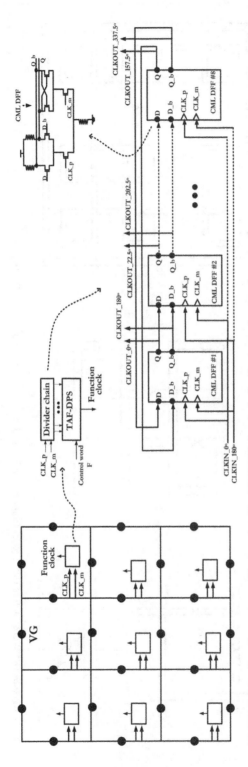

FIGURE 6.7 Standing wave oscillator assisted by TAF-DPS for clock distribution and generation.

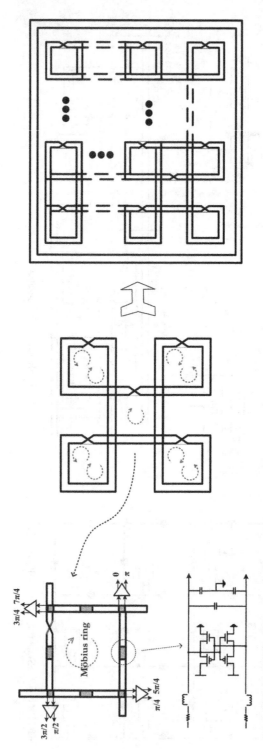

FIGURE 6.8 RTWO structure (left), ROA (middle), and chip size clock distribution using ROA (right)

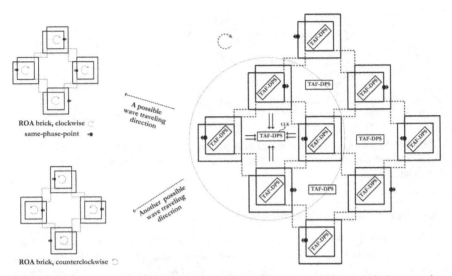

FIGURE 6.9   ROA brick (left); chip size clock distribution using ROA and clock generation using TAF-DPS (right).

determined by their physical positions. With each region surrounded by the lines, a TAF-DPS can be created to generate the function clock. All the outputs from the TAF-DPSs can have the same frequency and same phase, which satisfies the requirement of clock distribution for synchronous operation. Following is a summary of RTWO and TAP-DPS for clock distribution and clock generation:

- The same phase point of the whole chip is identifiable by connecting RTWO rings in a ROA brick fashion.
- The sequential order of the multiple phases is established from the same phase point and the physical positions of the taper out points.
- The TAF-DPS clock generator of each region is created from the multiple phases generated from each RTWO ring. Its output is used as a function clock to support local operation.
- The TAF-DPS outputs of all the regions are synchronous.
- Frequency synthesis is accomplished by TAF-DPS.

6.3   TAF-DPS CLOCK SOURCE AND FPGA

The FPGA (field programmable logic array) emerged in the 1980s as an alternative to the programmable logic device and application-specific integrated circuits (ASICs). It offers the significant benefit of being readily programmable. Unlike ASICs, it can be programmed multiple times in the field, giving designers valuable opportunities to tweak their circuits.

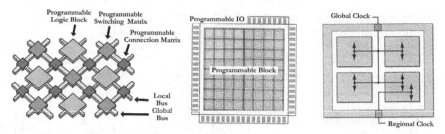

FIGURE 6.10   Programmable cell and programmable interconnect in FPGA (left); FPGA chip is made of programmable blocks and programmable IOs (middle); global and regional clocks (left).

6.3.1   FPGA Operation and Flexibility

The basic building elements in FPGA are wires, logic cells, and storage cells. Structurally, as illustrated in Figure 6.10, an FPGA chip consists of an array of logic blocks (made of logic and storage cells) that are configured by software. Those logic blocks are surrounded by programmable input–output (IO) blocks. Both the logic blocks and the IO blocks are connected by programmable interconnects (wires). The programming technology includes antifuse, SRAM, and flash. It determines the type of logic cells and the interconnect scheme. The modern FPGA can go further in complexity by including many hard IPs (intellectual property), such as memory controller, PCIe controller, clock generator, big block of memory, and even processor cores (e.g., ARM core). Those on-chip IPs significantly increase FPGA functionality and, at the same time, reduce user design effort.

Compared to conventional chip design approaches, what makes the FPGA unique is flexibility. The function achieved by an FPGA chip can be slightly modified or completely changed to accommodate various application scenarios. The trade-off between power and performance can be frequently updated for various needs. This inherent flexibility in the FPGA is naturally in agreement with the goal of TAF-DPS clock generator. A flexible frequency source enabled by TAF-DPS can further enhance FPGA capability. This is because, in principle, the frequency requirement from a user of a FPGA chip is unpredictable. As a result, the FGPA is preferred to possess the capability of providing any asked frequency. From the operation perspective, structurewise, the storage cells used in FPGA are all digital elements. A time-average-frequency clock signal can be used straightforwardly in driving them. In other words, the complex spectrum content of a TAF clock is not a concern in this application.

6.3.2   Prefabricated FPGA Structure and Clock/Frequency Distribution Requirement

Physically, unlike ASIC, which is customarily designed for each case and for each application, FPGA is premanufactured with a fixed structure for each family. An FPGA chip from a particular family can be used for multiple applications. This characteristic makes FPGA design unique, especially from a clocking perspective.

As illustrated in Figure 6.10, the preassembled cells (including large IPs, which are not shown in the drawing) are physically located in their assigned locations. Thus, it requires a plan for distributing the clock signal. Unlike the ASIC, this clock distribution plan must be generic since the FPGA is intended to serve many different applications. The uniqueness of its clock/frequency requirement can be summarized as follows:

- The clock signal frequency is preferred to be flexible since a FPGA chip needs to support a variety of applications that are unknown at the FPGA design time.
- There is a need for the global clock signal to be distributed to all the areas of the chip since globally synchronous operation must be supported.
- It is likely that different areas inside the FPGA need different clocking support. In other words, besides a global clock, there could be many regional clocks that all have unique characteristics.

6.3.3   Heterogeneous FPGA Computing and Frequency Requirement

A modern FPGA chip can include many different kinds of processing cores, such as CPU, GPU, and DSP, to support a variety of tasks, such as general-purpose processing, graphic-oriented processing, and digital signal processing. This heterogeneous computing demands many operating frequencies that must be generated from inside the FPGA. Further, communication among the processing cores, memory, and other IPs requires small frequency granularity and fast frequency switching. This fact calls for a powerful on-chip frequency/clock generator.

6.3.4   Parallelism in FPGA and Frequency Requirement

A unique feature of the FPGA is its capability to decompose a particular task into many smaller subtasks that can be subsequently executed in parallel. This is possible because a FPGA is made of many cells of the same structure. This parallelism makes it possible to achieve higher performance and lower power consumption in some applications. On this front, a flexible frequency source (such as TAF-DPS) can certainly enhance FPGA capability in partitioning a task into many subtasks. It provides an additional degree of freedom.

6.3.5   TAF-DPS Clock Source and FPGA

Commercial FPGA chips have complex clock generators built inside its structure [Alt12, Xil13]. These clock generators mainly perform the following functions:

- Frequency synthesis, to generate appropriate frequencies for functions
- Delay compensation, to align data and clock (zero delay, deskew)
- Tracking and cleaning of external clock signal
- Special clock function generation, such as spread spectrum clock generation

The circuitries used for achieving these functions are traditional charge pump PLL and DLL. The VCO or delay line used in PLL/DLL is a delay-buffer-based ring or chain. Usually, multiple outputs are available from the PLL/DLL for performing compensation related to data-clock alignment. In some cases, an internally generated signal can also be used as PLL input for phase alignment. Frequency synthesis is carried out using PLL plus prescale and postdivision counters. FPGA vendors use these building blocks in a sophisticated way to achieve the various functions discussed above. A special clock management element [such as the so-called clock management tile (CMT)] is created from these building blocks. They are placed in multiple locations of the chip to support operations that must be carried out in the corresponding regions.

Since a TAF-DPS clock generator has both the frequency synthesis and delay compensation capabilities (phase synthesis), inside a FPGA, it is a good candidate for performing the function mentioned above. Its frequency generation capability (illustrated in Figure 3.5) and delay compensation capability (illustrated in Figure 6.4) can be made available from a core cell in CMT (to replace the PLL and the DCM). The advantages of this TAF-DPS based FPGA clocking method can be summarized as follows:

- Small frequency granularity (sub-ppm) is possible from the FPGA.
- Dynamical frequency adjustment (fast frequency switching within the application process) can be possible. In other words, frequency can be treated as a programmable variable.
- The user interface for clock management can be much simplified and thus be FPGA user friendly.

6.3.6 Creating Clock Source of Sub-ppm Frequency Granularity and Two-Cycle Frequency Switching Speed with FPGA Custom Logic

In any FPGA chip, an essential element is the storage cell. A typical FPGA can have a large amount of on-chip flip-flops. Furthermore, the flip-flops can be driven by an external clock of relatively high frequency (hundreds of megahertz). Therefore, by chaining multiple flip-flops together (such as the Johnson counter) and driving the chain with an external high-frequency clock, multiple phases can be created at a lower frequency (refer to Figure 5.12 for an example). These phases can be used to help create a flexible TAF-DPS inside the FPGA. This TAF-DPS is made completely of logic gates that are available from the FPGA. Based on need, it can be programmed and be tweaked many times in the field. The fractional part of the frequency control word can be made arbitrarily large for fine frequency resolution. This is a convenient way to create a clock source having small frequency granularity, such as sub-ppm, for certain applications. Furthermore, the FPGA can be readily reprogrammed in the field for different frequency granularities.

Inside a FPGA, the TAF-DPS can be supported by another resource. Usually, a FPGA has one or several multiple-phase PLLs or DLLs on-chip. Therefore, it is possible to use such multiple outputs to build a TAF-DPS. This scheme can serve the goal of making a low-cost small-frequency-granularity dynamic clock source in the field as well.

REFERENCES

[Alt12]   Altera "Clock network and PLLs in the Cyclone III device family", *Cyclone III Device Handbook*, Vol. 1, Altera, Chapter 5. 2012. http://www.altera.com/literature/hb/cyc3/cyc3_ciii51006.pdf

[Cha03]   S. C. Chan, K. L. Shepard, and P. J. Restle., "Design of resonant global clock distributions", *Proc. Int. Conf. Computer Des.*, pp. 248–253, Oct. 2003.

[Cha04]   S. Chan, P. Restle, K. Shepard, N. James, and R. Franch, "A 4.6GHz resonant global clock distribution network", *Proc. IEEE Int. Solid-State Circuits Conf.*, vol. 1, pp. 342–343, Feb. 2004.

[Cha09]   S. Chan et al., "A resonant global clock distribution for the cell broadband engine processor", *IEEE Solid-State Circuits*, vol. 44, no. 1, pp. 64–72, Jan. 2009.

[Chu04]   J.-Y. Chueh, M. C. Papaefthymiou, and C. H. Ziesler, "Two-phase resonant clock distribution," *Proc. IEEE Computer Soc. Ann. Symp.*, VLSI, 2005.

[Fra03]   F. O'Mahony, C. P. Yue, M. A. Horowitz, and S. S. Wong, "A 10-GHz global clock distribution using coupled standing-wave oscillators," *IEEE J. Solid-State Circuits*, vol. 38, no. 11, pp. 1813–1820, Nov. 2003.

[Gut00]   V. Gutnik and A. P. Chandrakasan, "Active GHz clock network using distributed PLLs," *IEEE J. Solid-State Circuits*, vol. 35, no. 11, pp. 1553–1560, 2000).

[Mam09]   M. Sasaki, "A high-frequency clock distribution network using inductively loaded standing-wave oscillators," *IEEE J. Solid-State Circuits*, vol. 44, no. 10, pp. 2800–2807, Oct. 2009.

[Pra95]   G. A. Pratt et al., "Distributed Synchronous clocking", *IEEE Trans. Parallel Distributed Sys.*, vol. 6, no. 3, pp. 314–328, Mar. 1995.

[Sat08]   V. S. Sathe, J. C. Kao, and M. C. Papaefthymiou, "Resonant-clock latch-based design," *IEEE J. Solid-State Circuits*, vol. 43, no. 4, pp. 864–873, Apr. 2008.

[Tam09]   S. Tam et al., "Clock generation and distribution for a 45 nm, 8-core Xeon® processor with 24 MB," *Symp. VLSI Circuits Dig. Tech. Pap.*, June 2009.

[Ten11]   Y. Teng, J. Lu and B. Taskin, "ROA-brick topology for rotary resonant clocks," Paper presented at the 2011 IEEE 29th International Conference on Computer Design (ICCD), pp. 273–278, 2011.

[Woo01]   J. Wood, T. C. Edwards, and S. Lipa, "Rotary traveling-wave oscillator arrays: A new clock technology," *IEEE J. Solid-State Circuit*, Vol. 36, no. 11, pp. 1654–1650, Nov. 2001.

[Xil13]   Xilinx, *Spartan-6 FPGA Clocking Resources*, User Guide, Xilinx, 2013. http://www.xilinx.com/support/documentation/user_guides/ug382.pdf

[Xiu12]   L. Xiu, *Nanometer Frequency Synthesis beyond Phase Locked Loop*, Wiley-IEEE Press, Hoboken, NJ, Aug. 2012.

[Zia13]   E. Zianbetov, et al., "Distributed clock generator for synchronous SoC using ADPLL network", Paper presented Custom Integrated Circuits Conference (CICC), IEEE, pp. 1-4, 2013.

[Zha09]   Y. Zhang, J. F. Buckwalter, and C.-K. Cheng, "On-chip global clock distribution using directional rotary traveling-wave oscillator," Paper presented at the Conference on IEEE 18th Electrical Performance of Electronic Packaging and Systems, EPEPS '09, 2009, pp. 251–254.

7

DIGITAL-TO-FREQUENCY CONVERTER: A COMPONENT ENABLING NEW APPLICATION FIELDS

The TAF-DPS's capability of arbitrary frequency generation (small frequency granularity) and instantaneous frequency switching (two cycles) makes it a new type of signal converter: digital-to-frequency converter (DFC). The process of digital-to-frequency conversion is similar to the process of digital-to-analog conversion. They are illustrated in Figure 7.1 side by side. The difference is at the outputs. The DAC output is a signal waveform where the voltage value at each sample point is the focus point. In contrast, the DFC output is a pulse train whose frequency (pulse length, period) at each individual period is the intended information. Table 7.1 compares the DFC and the DAC from multiple perspectives.

There are several architectures in DAC design: oversampling DAC, binary-weighted DAC, R-2R ladder DAC, and thermometer-coded DAC. They are all based on the *additive principle* since multiple voltage/current values can be added together to form new information. The DFC, on the other hand, is less complex in terms of its working principle. It simply counts a base time unit to create each pulse's length-in-time (the period and therefore the frequency). Unlike voltage and current, which are representations of certain amounts of electrical charge, frequency (time) is not a physical entity and is not additive (by contrast, electrical charges can be summed). For example, when a frequency source of 10 MHz is "added" with another frequency source of 100 MHz, the result is not simply a frequency of 110 MHz. It is a nonlinear operation. The "add" operation is non-existent for frequency (time) sources. For this reason, there is only one practical architecture for DFC: Directly

From Frequency to Time-Average-Frequency: A Paradigm Shift in the Design of Electronic Systems, First Edition. Liming Xiu.
© 2015 The Institute of Electrical and Electronics Engineers, Inc. Published 2015 by John Wiley & Sons, Inc.

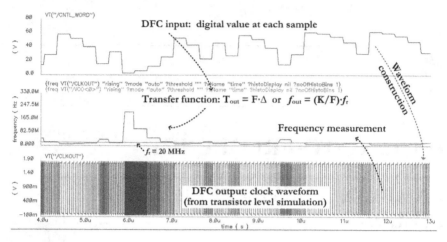

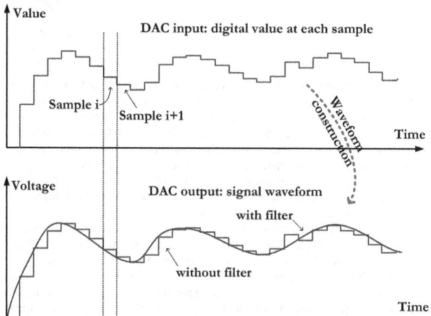

FIGURE 7.1   Digital-to-frequency conversion process (top) and digital-to-analog conversion process (bottom).

construct each individual pulse by controlling its low and high time durations (thus the term *direct period synthesis*).

Because of the additive principle and the time-sampling mechanism, the figure of merit for DAC performance is rather complicated. It includes voltage resolution, gain, offset, noise, maximum sampling rate, monotonicity, dynamic range, total harmonic distortion and noise (THD+N), differential nonlinearity (DNL), integral

TABLE 7.1   Comparison of DAC and DFC (TAF-DPS)

	DAC	DFC (TAF-DPS)
Input	Digital value	Digital value
Output	Analog voltage	Frequency, time if a counter is attached afterward
Reference input	Reference voltage	Reference frequency
Transfer function	$V_{out} = CF_V^*$	$f_{out} = (K/F)f_r$
Output range	VSS to VDD (or smaller)	$\frac{1}{2}f_r^a \leq f_{out} \leq (K/2)f_r$
Working principle	Voltage and current are additive	Time-average-frequency
Based unit	Small precise "unit voltage/current"	Small precise "unit time"
Resolution	Voltage resolution: unit voltage	Time resolution Δ, frequency resolution $df/f = dF/F$
Switching speed	Slew rate (load dependent)	Two cycles

Note: F_V = voltage control word; F = frequency control word.
[a]Low side can be expanded by a frequency divider.

nonlinearity (INL), spurious free dynamic range (SFDR), signal-to-noise-and-distortion ratio (SNDR), ith harmonic distortion (HDi), and total harmonic distortion (THD). In contrast, there are fewer figures of merit for DFC since only "time," excluding voltage and current, is the design parameter of interest. Specially, because the additive principle is not involved, monotonicity is guaranteed for DFC. This is true because a counter's output always grows when its index number increases. For DFC, the key figures of merit are frequency/time resolution (frequency/time granularity) and frequency switching speed. These two parameters are directly related to the measure of how many events happened in a given time window (rate-of-switching). Additionally, if DFC is used as a clock signal generator or if its output spectrum is of significance, jitter/phase noise and spectrum purity will be other concerns.

DAC's major applications are audio and video. In these applications, audio/video information is originally processed in the digital domain where high precision (high fidelity) can be preserved during processing and transmission. Only at the last stage is audio/video information converted to analog voltage by DAC so that it can be heard/viewed by us. Although DFC is based on TAF-DPS, its purpose is not just for frequency synthesis (clock generation). For example, it can be used as an on–off switch to control the flow of certain physical materials such as fluid, gas, light, electrical charge, and magnetic flux. Thus, DFC can also be taken as digital-to-flow converter. This type of converter uses the high/low percentage in voltage level (refer to the three types of TAF-PWM outputs in Figures 5.49 and 5.50) to control the amount of the intended media. It can function as charge pump, fluid pump, flux pump, and light pump in various applications. It can be very useful in industry control, especially when the control mechanism is required to be carried out at high speed (hundreds of megahertz) and high precision (parts per million). Currently, these types of controls are realizable using a pulse-counter-based digital circuit. As a result, the control is coarse since it is based on the mechanism of counting the number of clock

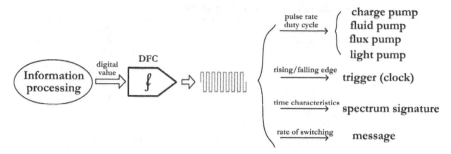

FIGURE 7.2   Potential applications of DFC.

cycles. Additionally, each cycle is fixed (untouchable). In the DFC case, each pulse (cycle) length is controllable. The pulse-counter-based control mechanism is also relatively slow since it divides down the reference frequency. Figure 7.2 illustrates DFC's potential applications. In addition to flow control and clocking, DFC can produce a unique spectrum signature for each digital value. This could be useful in certain cases.

Figure 7.3 illustrates an information processing flow based on "rate-of-switching." In this approach, the sensor outputs a pulse train which can be in the form of voltage, current, sound, light, etc. Instead of the voltage or current level as in the case of ADC, the intended information in this case is the number of "zero-crossing" points within a given time window. The zero-crossing represents the rate-of-switching of certain physical phenomena being monitored by the sensor. This information can be processed by an FDC (frequency-to-digital converter) and be converted into a digital value. The digital signal processing unit then further processes the information and produces the desired output in digital format. On the actuating side, the DFC can be used to carry out the desirable action by controlling the rate-of-switching of the pump (digital pump and valve). For certain physical phenomena, this rate-of-switching information processing flow can be more cost-effective and/or more precise. This is because many real world events are simply vibrational. In these cases, the speed of vibration is what we care about. Therefore, there is no need to convert this vibration information back and forth (to voltage level and then back to electrical oscillation). In the architecture of Figure 7.3, there are four types of converter,

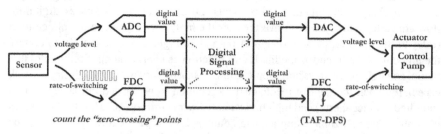

FIGURE 7.3   Information processing flow based on rate-of-switching.

ADC, FDC, DAC and DFC, and two types of information formats, level and rate-of-switching. Depending on the problem being investigated, they can be used in various combinations.

DFC is a newcomer to the converter family (it has been the missing member of the converter family for a long time) and it is the enabling component for rate-of-switching information processing flow.

8

CONCLUSION: EVOLUTION OF TIMEKEEPER IN ELECTRICAL WORLD

Among the millions of signals in a microelectronic system, the clock signal is one that oscillates between a high and a low state and is utilized as a metronome to coordinate the actions/events of the system. It is the timekeeper in the electrical world. In the early days, a clock signal generator was only required to produce a pulse train of fixed frequency. This was sufficient for supporting system operation at that time. Among all types of single-frequency clock generators, they all share the basic parts of a resonant circuit and an amplifier. The resonant circuit is responsible for originating the oscillation and producing the pulse train. It can be a quartz oscillator, an *LC* tank circuit, or an *RC* oscillating circuit. The amplifier circuit inverts the signal from the resonant circuit (the oscillator) and feeds a portion of it back to the oscillator to maintain the oscillation.

As system complexity grows, the single-frequency clock generator can no longer fulfill the needs. The frequency divider and frequency multiplier are developed to generate more clock frequencies. The frequency divider includes analog dividers (e.g., regenerative frequency divider and injection-locked frequency divider) and digital divider (flip-flop based digital counter). Direct digital synthesis (DDS) can also be viewed as a frequency divider since it derives its output frequency from a high-frequency reference. Frequency multiplication is achieved by using phase-locked loops (integer-N PLL and fractional-N PLL). The divider and multiplier allow a variety of frequencies to be generated and thus potentially selected for system use without modifying the hardware.

From Frequency to Time-Average-Frequency: A Paradigm Shift in the Design of Electronic Systems, First Edition. Liming Xiu.
© 2015 The Institute of Electrical and Electronics Engineers, Inc. Published 2015 by John Wiley & Sons, Inc.

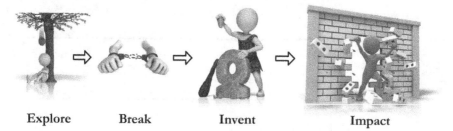

Explore   **Break**   **Invent**   **Impact**

FIGURE 8.1 The process of making advancement in engineering. (original images are provided by PresenterMedia)

However, the frequency divider and multiplier have not solved the problems emerging from modern complex system design: *the need of arbitrary frequency generation and instantaneous frequency switching.* The solution to the first problem requires a reinvestigation into the concept of frequency. This effort is beyond simply playing with various techniques at the circuit level. Philosophically, it requires an adjustment in our thinking about the clocking of electrical devices. This probing leads to the use of time-average-frequency. The cure for the second problem demands the adoption of a direct approach. The combined efforts enable the new frequency synthesis method of time-average-frequency direct period synthesis, or TAF-DPS for short.

At this level of frequency generation capability, TAF-DPS aims at the target of becoming a field programmable frequency generator (FPFG). It sends the following messages to microelectronic system designers:

- Frequency is an entity that is programmable, just like the microprocessor instruction set in the hand of a software programmer.
- Frequency is an entity that can directly convey information, just like voltage in the hand of an analog circuit designer.

The introduction of Time-Average-Frequency concept and theory and the invention of TAF-DPS follow the flow of exploring new idea, breaking conventional wisdom, inventing new stuffs and then impacting a field. This flow of making advancement in engineering is illustrated in Figure 8.1. It is this author's wish that more researchers and engineers would be inspired and subsequently join this effort.

INDEX

From Frequency to Time-Average-Frequency: A Paradigm Shift in the Design of Electronic Systems, First Edition. Liming Xiu.
© 2015 The Institute of Electrical and Electronics Engineers, Inc. Published 2015 by John Wiley & Sons, Inc.

IEEE PRESS SERIES ON MICROELECTRONIC SYSTEMS

The focus of the series is on all aspects of solid-state circuits and systems including the design, testing, and application of circuits and subsystems, as well as closely related topics in device technology and circuit theory. The series also focuses on scientific, technical and industrial applications, in addition to other activities that contribute to the moving the area of microelectronics forward.

R. Jacob Baker, *Series Editor*